Topics in Graph Theory

Topics in Graph Theory

Graphs and Their Cartesian Product

Wilfried Imrich
Sandi Klavžar
Douglas F. Rall

 CRC Press
Taylor & Francis Group
Boca Raton London New York

CRC Press is an imprint of the
Taylor & Francis Group, an **informa** business

AN A K PETERS BOOK

First published 2008 by A K Peters, Ltd.

Published 2019 by CRC Press
Taylor & Francis Group
6000 Broken Sound Parkway NW, Suite 300
Boca Raton, FL 33487-2742

© 2008 by Taylor & Francis Group, LLC
CRC Press is an imprint of Taylor & Francis Group, an Informa business

First issued in paperback 2019

No claim to original U.S. Government works

ISBN 13: 978-0-367-44610-9 (pbk)
ISBN 13: 978-1-56881-429-2 (hbk)

Visit the Taylor & Francis Web site at
http://www.taylorandfrancis.com

and the CRC Press Web site at
http://www.crcpress.com

Library of Congress Cataloging-in-Publication Data

Imrich, Wilfried, 1941-
    Topics in graph theory : graphs and their cartesian product / Wilfried Imrich, Sandi Klavžar, Douglas F. Rall.
        p. cm.
    Includes bibliographical references and index.
    ISBN 978-1-56881-429-2 (alk. paper)
    1. Graph theory. I. Klavžar, Sandi, 1962- II. Rall, Douglas F. III. Title.

QA166.I469 2008
511'.5-dc22

2008020790

To our wives Gabi, Maja, and Naomi.
Without their love, patience, encouragement, support, and
understanding, the chances of this book being published
would have been infinitesimal at best.

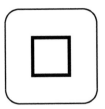

Contents

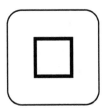

Preface

Graphs have become a convenient, practical, and efficient tool to model real-world problems. Their increasing utilization has become commonplace in the natural and social sciences, in computer science, and in engineering. The development of large-scale communication and computer networks as well as the efforts in biology to analyze the enormous amount of data arising from the human genome project are but two examples.

Not surprisingly, courses in graph theory have become part of the undergraduate curriculum of many applied sciences, computer science, and pure mathematics courses. Due to the complexity of the applications, many graduate programs in these areas now include a study of graph theory.

A multitude of excellent introductory and more advanced textbooks are on the market. In this book, we address a reader who has been exposed to a first course in graph theory, wishes to apply graph theory at a higher or more special level, and looks for a book that repeats the essentials in a new setting, with new perspectives and results. For this reader, we wish to communicate a working understanding of graph theory and general mathematical tools. The prerequisites are previous exposure to fundamental notions of graph theory, discrete mathematics, and algebra. Therefore, we will not strain the reader's patience with definitions of concepts such as equivalence relations or groups.

The context we chose for this task are graph products and their subgraphs. This includes Hamming graphs, prisms, and many other classes of graphs that are either graph products themselves or are closely related to them—often in surprising, unexpected ways.

This setting allows us to cover concepts with applications in many fields of mathematics and computer science. It includes problems from coding theory, frequency assignment, and mathematical chemistry, which are briefly treated to give the reader a flavor of the variety of the applications.

Many results in this book are recent in the sense that they first appeared in print around the time this book went to press. We have taken efforts to present them accurately and efficiently in a unified environment.

The book is divided into five parts. The first part is a short introduction to the Cartesian product—the main tool that is used throughout the remainder of the book. We convey the basic facts about the product, and apply them to Hamming graphs and Tower of Hanoi graphs, that is, to two classes of graphs that naturally appear.

Classic topics of graph theory are treated in Part II. Included are the fundamental notions of hamiltonicity, planarity, connectivity, and subgraphs. These standard concepts are introduced in most typical first courses in graph theory. We include several interesting results about these basic concepts, which were, somewhat surprisingly, only recently proved. Nonetheless, many challenging open problems still exist in these areas. For example, there is the unsettled conjecture by Rosenfeldt and Barnette that the prism over a 3-connected planar graph is hamiltonian and the determination of the crossing number of the so-called "torus graphs."

A large part of graph theory involves the computation of graphical invariants. The reason is that many applications in different fields reduce to such computations. It turns out that a variety of scheduling and optimization problems are actually coloring problems in graphs constructed from the constraints. In Part III, we therefore focus on several different graph coloring invariants, some standard and some more recently introduced. In a separate chapter we study the problem of determining the cardinality of a largest independent set in a graph. The remaining two chapters of Part III focus on the domination number of a graph with special emphasis on the famous conjecture of Vizing.

Distances in graphs represent another major area for applications. As an example of such an application we present the Wiener index, which is probably the most explored topological index in mathematical chemistry. In Part IV, we demonstrate that the Cartesian product is a natural environment for the standard shortest-path metric. The starting point for this is the fact that the distance function is additive on product graphs. The material in this part of the book culminates in the Graham-Winkler Theorem, asserting that every connected graph has a unique canonical, isometric embedding into a Cartesian product.

Mathematical structures can be properly understood only if one has a grasp of their symmetries. It also helps to know whether they can be constructed from smaller constituents. This approach is taken in Part V. It leads to the prime factorization of graphs and the description of their automorphism groups. These, in turn, simplify the investigation of algebraic properties of connected or disconnected graphs with respect to the Cartesian product. In particular, cancelation properties are derived and the unique r^{th} root property is proved. Thereafter follows a chapter on the recent concept of the distinguishing number, which measures the effort needed to break all symmetries in a graph. The last chapter shows how the main result on the structure and the symmetries of Cartesian products lead to efficient factorization algorithms and the recognition of partial cubes.

Every chapter ends with a list of exercises. They are an integral part of the book because we are convinced that problem solving is not only at the core of mathematics, but is also essential for the comprehension and acquisition of mathematical proficiency. Checking one's mastery of ideas is crucial for strengthening self-confidence and self-reliance. Therefore some of the exercises are computational; others ask for the proof of a result in the chapter. The easier exercises let the reader check whether he or she grasps the concepts, but most of the exercises require an original idea, and a few demand a higher level of abstraction. Then there are problems whose solution requires the investigation of numerous cases. The idea for these problems is to find a way to minimize the effort and to solve some of the cases.

Hints and solutions to the exercises are collected at the end of the book.

We cordially thank Drago Bokal, Mietek Borowiecki, Boštjan Brešar, Ivan Gutman, Bert Hartnell, Iztok Peterin, and Simon Špacapan for invaluable comments, remarks and contributions to the manuscript. We are especially grateful to Amir Barghi, a graduate student at Dartmouth College, for a careful reading of the entire manuscript. His suggestions led to improvements in the presentation at numerous places in the text.

The manuscript was tested in courses at the University of Maribor, Slovenia; the Montanuniversität Leoben, Austria; and Furman University, Greenville, SC, United States. We wish to thank our students Matevž Črepnjak, Michael Hull, Marko Jakovac, Luka Komovec, Aneta Macura, Michał Mrzygłód, Mateusz Olejarka, Katja Prnaver, Jeannie Tanner, and Joseph Tenini for remarks that helped to make the text more accessible.

Last, and certainly not least, we wish to thank Charlotte Henderson, our associate editor, and the other staff at A K Peters, Ltd., for the professional support and handling of our book that every author desires. Special thanks are extended to Alice and Klaus Peters for their involvement and expertise offered at all stages of publication.

W. Imrich, S. Klavžar, and D.F. Rall
Leoben, Maribor, Greenville
April 2008

Part I

Cartesian Products

1   The Cartesian Product

Throughout this book the Cartesian product will be the leading actor. With its help, the reader will develop a deeper understanding of graph theory. In addition, the reader will learn about important new concepts such as circular colorings, $L(2,1)$-labelings, prime factorizations, canonical metric embeddings, and distinguishing numbers.

In this chapter, we define the Cartesian product and introduce fibers and projections as important tools for further investigations. We also show that a product graph is connected if and only if its factors are connected. Along the way, we list several examples of Cartesian products. In particular, we observe that line graphs of complete bipartite graphs are products of complete graphs, and we show that these are the only products that are line graphs.

1.1   Definitions, Fibers, and Projections

Before we define the Cartesian product, we list some conventions to be used throughout the book. We write $g \in G$ instead of $g \in V(G)$ to indicate that g is a vertex of G, and $|G| = |V(G)|$ for the number of vertices. An edge $\{u, v\}$ of a graph G is denoted as uv. Sometimes, particularly when dealing with edges in products, we also write $[u, v]$.

The *Cartesian product* of two graphs G and H, denoted $G \square H$, is a graph with vertex set

$$V(G \square H) = V(G) \times V(H), \tag{1.1}$$

that is, the set $\{(g, h) \mid g \in G, h \in H\}$.

The edge set of $G \square H$ consists of all pairs $[(g_1, h_1), (g_2, h_2)]$ of vertices with $[g_1, g_2] \in E(G)$ and $h_1 = h_2$, or $g_1 = g_2$ and $[h_1, h_2] \in E(H)$.

For example, Figure 1.1 depicts $P_4 \square P_3$ (left) and $C_7 \square K_2$ (right). To the first example, we remark that Cartesian products $P_m \square P_n$ of two

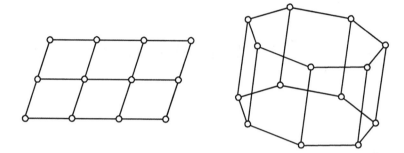

Figure 1.1. Cartesian products $P_4 \square P_3$ (left) and $C_7 \square K_2$ (right).

paths on m and n vertices are called *complete grid graphs*, and their subgraphs are known as *grid graphs*. Such graphs appear in many applications, for instance in the theory of communication networks.

Note that $K_2 \square K_2 = C_4$, that is, the Cartesian product of two edges is a square. This is the motivation for the introduction of the notation $\square$ for the Cartesian product.[1]

We can also define the edge set by the relation

$$E(G \square H) = (E(G) \times V(H)) \cup (V(G) \times E(H)), \tag{1.2}$$

where the edge (e, h) of $G \square H$, with $e = [g_1, g_2] \in E(G)$, $h \in H$, has the endpoints (g_1, h), (g_2, h), and the edge (g, f), with $g \in G$, $f = [h_1, h_2] \in E(H)$, has the endpoints (g, h_1), (g, h_2).

Since edges in simple graphs can be identified with their (unordered) sets of endpoints, the preceding two definitions of the edge set of $G \square H$ are equivalent.

Combining Equations (1.1) and (1.2), we obtain yet another, even more concise characterization of the Cartesian product of two graphs G and H; see Gross and Yellen [49, p. 238]:

$$G \square H = (G \times V(H)) \cup (V(G) \times H).$$

Here

$$G \times V(H) = \bigcup_{h \in H} (G \times \{h\}),$$

and every $G \times \{h\}$ is a copy of G. We denote it by G^h and call it a *G-fiber*.[2] Analogously, $V(G) \times H$ is the union of the *H-fibers* $^gH = \{g\} \times H$.

---

[1] Some authors use the term *box product* for the Cartesian product.
[2] In *Product Graphs* [66], fibers are referred to as *layers*.

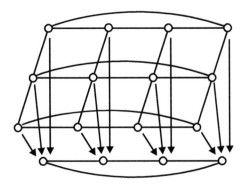

Figure 1.2. Projection $p_{C_4} : C_4 \,\square\, P_3 \to C_4$.

For a given vertex $v = (g, h)$, we also write $G^v = G^h$, and respectively, $^vH = {}^gH$. Clearly G^v can also be defined as the subgraph of $G \,\square\, H$ induced by $\{(x, h) \mid x \in G\}$, and the mapping $\varphi : V(G^h) \to V(G)$ defined by

$$\varphi : (x, h) \mapsto x$$

is a bijection that preserves adjacency and nonadjacency. Such a bijection is called a *graph isomorphism*. One can say G^h and G are isomorphic, or in symbols, $G^h \cong G$.

Sometimes we also write $G = H$ for isomorphic graphs. Thus, $G = H$ may mean that G and H have the same vertex and edge sets or that G and H are isomorphic. For example, the phrase "G is a K_2" or the equation "$G = K_2$" both mean that G is isomorphic to K_2.

In contrast, for two G-fibers G^v and G^w of a product $G \,\square\, H$, the statement $G^v = G^w$ expresses that these fibers are identical, whereas $G^v \cong {}^wH$ really means only that the two fibers are isomorphic because $G^v \neq {}^wH$; in fact, $|G^v \cap {}^wH| = 1$.

Note that $G \,\square\, H \cong H \,\square\, G$, and that $K_1 \,\square\, G \cong G \,\square\, K_1 \cong G$. In other words, Cartesian multiplication is commutative and K_1 is a unit (see Exercise 3).

For a product $G \,\square\, H$, the *projection* $p_G : G \,\square\, H \to G$ is defined by

$$p_G : (g, h) \mapsto g.$$

It is clear what we mean by p_H. See Figure 1.2 for the projection p_{C_4} of $C_4 \,\square\, P_3$ onto C_4.

Under the projections p_G or p_H, the image of an edge is an edge or a single vertex. Such mappings are called *weak homomorphisms*. Clearly, the mapping $\varphi : V(G^h) \to V(G)$ defined above is the restriction of p_G to G^h.

More generally, let U be a subgraph of $G \square H$. We follow common practice and define $p_G U$ as the subgraph of G with the vertex set

$$\{p_G(v) \mid v \in U\}$$

and edge set

$$\{[p_G(u), p_G(v)] \mid [u, v] \in E(U), p_G(u) \neq p_G(v)\}.$$

1.2   Connectedness and More Examples

We continue with the following simple, yet fundamental observation about Cartesian products.

Lemma 1.1. *A Cartesian product $G \square H$ is connected if and only if both factors are connected.*

Proof: Suppose $G \square H$ is connected. We have to prove that both G and H are connected. Clearly, it suffices to prove it for G. Let g and g' be any two vertices of G, and let $h \in H$ be arbitrary. Then there is a path P in $G \square H$ from (g, h) to (g', h), and $p_G P \subseteq G$ contains a path from g to g'.

Conversely, assume that G and H are connected. We have to show that there is a path between any two arbitrarily chosen vertices (g, h) and (g', h') of $G \square H$. Let R be a g,g'-path and S an h,h'-path. Then

$$(R \times \{h\}) \cup (\{g'\} \times S)$$

is a $(g, h),(g', h')$-path; see Figure 1.3.   □

Before continuing with new concepts related to Cartesian products, we take a break with two examples: prisms and line graphs of complete bipartite graphs.

Prisms over graphs appear in many situations and are defined as follows. Let G be a graph. Then the *prism over G* is the graph obtained from the disjoint union of graphs G' and G'', both isomorphic to G, by joining any vertex of G' with its isomorphic image in G''. An example of a prism is shown in Figure 1.4.

From our point of view, the prism over G is the Cartesian product

$$G \square K_2.$$

Let G be a graph. Then the vertex set of the *line graph $L(G)$ of G* consists of the edges of G. Two vertices of $L(G)$ are adjacent if the

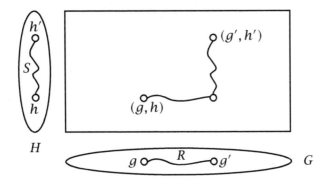

Figure 1.3. $G \square H$ is connected provided G and H are connected.

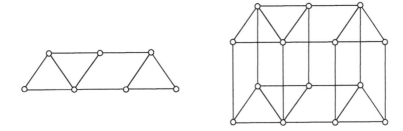

Figure 1.4. A graph (left) and the prism over it (right).

corresponding edges of G are adjacent. Note that a vertex of degree d in G yields a complete subgraph K_d in $L(G)$.

For instance, for any $n \geq 2$, $L(K_{1,n}) = K_n$ and $L(P_n) = P_{n-1}$.

Let $m, n \geq 2$ and consider the complete bipartite graph $K_{m,n}$ with the bipartition $\{u_1, u_2, \ldots, u_m\}$, $\{w_1, w_2, \ldots, w_n\}$. Then the vertex set of $L(K_{m,n})$ is

$$\{u_i w_j \mid 1 \leq i \leq m, 1 \leq j \leq n\}.$$

Vertices $u_i w_j$ and $u_{i'} w_{j'}$ are adjacent in $L(K_{m,n})$ if and only if $i = i'$ and $j \neq j'$, or $j = j'$ and $i \neq i'$. This implies that

$$L(K_{m,n}) = K_m \square K_n.$$

In fact, these are the only connected line graphs that are Cartesian products, as the following result of Palmer asserts.

Proposition 1.2. [97] *Let X be a connected graph. Then $L(X)$ is a nontrivial Cartesian product if and only if $X = K_{m,n}$, $m, n \geq 2$.*

Proof: We have already observed that the line graph of a complete bipartite graph is a Cartesian product.

Suppose $L(X) = G \square H$, where $2 \leq |H| \leq |G|$. Since X is connected, $L(X)$ is connected (see Exercise 14) and, by Lemma 1.1, its factors G and H are connected as well. If $|G| = 2$, then $G = H = K_2$ and there is nothing to show.

Hence, we can assume that G has at least three vertices. Since it is connected, it contains a path of length 2, say g_1, g_2, g_3. Consider an arbitrary edge $h_1 h_2$ of H. Then, in $G \square H$, the vertices (g_1, h_1), (g_2, h_1), (g_3, h_1), and (g_2, h_2) induce a $K_{1,3}$ unless $g_1 g_3 \in E(G)$. But $K_{1,3}$ is a forbidden induced subgraph in line graphs (see Exercise 15). Thus, $g_1 g_3$ is an edge of G. As G is connected, we infer by simple induction that G is a complete graph (see Exercise 16). For $|H| > 2$, an analogous argument shows that $|H|$ is also complete.

Finally, since the only nonisomorphic graphs with the same line graph are K_3 and $K_{1,3}$ (this is a result of Whitney [116] from 1932), we conclude that $X = K_{|G|,|H|}$. $\square$

1.3   Products of Several Factors

The Cartesian product is associative;[3] that is,

$$(G \square H) \square K \cong G \square (H \square K)$$

for arbitrary graphs G, H, and K (see Exercise 4). Hence, we do not need parentheses for products of more than two factors.

This allows an alternative direct definition of the Cartesian product of several factors. The *Cartesian product*

$$G = G_1 \square G_2 \square \cdots \square G_k$$

of the graphs $G_1, G_2, \ldots, G_k$ is defined on the k-tuples $(v_1, v_2, \ldots, v_k)$, where $v_i \in G_i$, $1 \leq i \leq k$. Two k-tuples $(u_1, u_2, \ldots, u_k)$ and $(v_1, v_2, \ldots, v_k)$ are adjacent if there exists an index ℓ such that $[u_\ell, v_\ell] \in E(G_\ell)$ and $u_i = v_i$ for $i \neq \ell$. The k-tuples $(v_1, v_2, \ldots, v_k)$ are called *coordinate vectors*, and the v_i are the *coordinates*.

In Figure 1.5, the Cartesian products $K_2 \square K_2 \square C_6$ and $P_3 \square P_5 \square P_6$ are shown.

The Cartesian product $G = G_1 \square G_2 \square \cdots \square G_k$ of several factors will be briefly denoted as

$$G = \square_{i=1}^{k} G_i .$$

---

[3]Since it is also commutative, and since K_1 as a unit, simple graphs form a commutative monoid with respect to Cartesian multiplication.

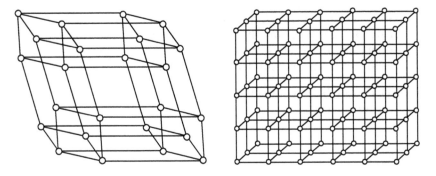

Figure 1.5. Cartesian products $K_2 \square K_2 \square C_6$ (left) and $P_3 \square P_5 \square P_6$ (right).

As in the case of two factors, $\square_{i=1}^{k} G_i$ is connected if and only if all factors G_i are connected. For a positive integer n, the nth Cartesian power of a graph G is $G^n = \square_{i=1}^{n} G$.

The projection $p_{G_i} : G \to G_i$ is defined by $p_{G_i} : (v_1, v_2, \ldots, v_k) \mapsto v_i$. Often we simply write p_i instead of p_{G_i}. Again, as in the case of two factors, the projections are weak homomorphisms.

1.4 Exercises

1. Determine the number of vertices of $G \square H$ in terms of G and H.

2. Determine the number of edges of $G \square H$ in terms of G and H.

3. Show that Cartesian multiplication is commutative, and that K_1 is a unit. (Hint: Show that the mapping $(g, h) \mapsto (h, g)$ is an isomorphism.)

4. Show that Cartesian multiplication is associative.

5. [63] Show that $K_3 \square K_3$ is isomorphic to its complement. (In other words, $K_3 \square K_3$ is a self-complementary graph.)

6. Let K be a square with one diagonal. Show that $K_2 \square K$ is self-complementary.

7. Show that the complement of $K_2 \square K_2 \square K_2$ is $K_4 \square K_2$.

8. Given a product $G \square H$, show that $|G^h \cap {}^g H| = 1$ for any $g \in G$ and $h \in H$.

9. Given a product $G \square H$, two fibers G^{h_1} and G^{h_2} are called *adjacent* if $h_1 h_2 \in E(H)$. Show that the edges between the two fibers are independent;[4] that is, no two of them share a vertex, and they define an isomorphism between G^{h_1} and G^{h_2}.

10. Let G and H be connected, nontrivial graphs. Show that $G \square H$ has no cut vertex.

11. Show that the Cartesian product of bipartite graphs is bipartite.

12. Show that for any $n \geq 3$, $L(C_n) = C_n$.

13. Show that if G is connected, then the line graph $L(G)$ is a tree if and only if G is a path.

14. Show that the line graph of a connected graph is connected.

15. Show that the line graph of a graph does not contain $K_{1,3}$ as an induced subgraph.

16. Let G be a connected graph with the property that any two vertices that can be connected by a walk of length two are adjacent. Show that G is complete.

---

[4]Such a set of edges is also called a *matching*.

2   Hamming Graphs and Hanoi Graphs

In this chapter, we introduce two important classes of graphs: hypercubes and Hamming graphs. The first class forms one of the central models for interconnection networks and parallel computing, whereas Hamming graphs[1] play an important role in metric graph theory and coding theory. These two classes also host many interesting and applicable classes of graphs as (induced, spanning, and isometric) subgraphs; for instance, median and quasi-median graphs, partial cubes, partial Hamming graphs, and Tower of Hanoi graphs, to list just a few.

In the first section of this chapter, we define and characterize hypercubes and Hamming graphs. In the second, we treat the generalized Tower of Hanoi puzzle and show that the graphs describing the moves of the puzzle are spanning subgraphs of Hamming graphs.

2.1   Hypercubes and Hamming Graphs

The *n-cube* Q_n, $n \geq 1$, is defined as the nth power of K_2 with respect to the Cartesian product. More precisely, we set $Q_1 = K_2$, and for $n \geq 2$,

$$Q_n = Q_{n-1} \,\square\, K_2 \,.$$

We use the term *hypercubes* for the set of all n-cubes.

If we set $V(K_2) = \{0, 1\}$, then the vertex set of Q_n consists of the n-tuples $(v_1, v_2, \ldots, v_n)$, where $v_i \in \{0, 1\}$, and two n-tuples $(u_1, u_2, \ldots, u_n)$, $(v_1, v_2, \ldots, v_n)$ are adjacent if they differ in exactly one coordinate; see Figure 2.1 for Q_4.

The figure shows also that $Q_4 = Q_3 \,\square\, K_2$. One of the Q_3-fibers is drawn with bold lines, the other with broken ones. The remain-

---

[1]Often, especially in coding theory, Hamming graphs refer to a slightly smaller class of graphs—Cartesian products of complete graphs in which all factors are of the same order.

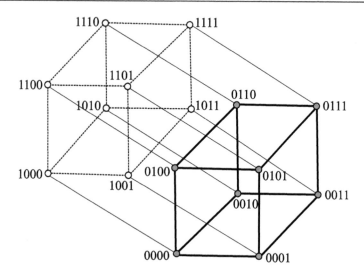

Figure 2.1. Graph of Q_4.

ing edges are the K_2-fibers of this presentation and induce an isomorphism between the Q_3-fibers.

Clearly, Q_n is an n-regular graph on 2^n vertices. Therefore, since the sum of the degrees in a graph is twice the number of its edges, Q_n contains $n2^{n-1}$ edges.

Hypercubes form one of the most studied classes of graphs, and it would be difficult to find some graph problem on hypercubes that has not been studied. As it turns out, despite the relatively simple structure of these graphs (after all, they are the simplest possible Cartesian product graphs with more than two factors), many problems on them are strikingly difficult. For instance, this is the case for the domination number, as we will see in Chapter 11.

Not surprisingly, many characterizations of hypercubes are known. As an example, we present one of Mulder [94]. Call a connected graph G a *(0, 2)-graph* if any two distinct vertices in G have exactly two common neighbors or none at all. Then Mulder proved:

Theorem 2.1. [94] *Let G be an n-regular $(0, 2)$-graph. Then $G = Q_n$ if and only if $|G| = 2^n$.*

A natural generalization of hypercubes are *Hamming graphs*, which are defined as Cartesian products of complete graphs. (In fact, hypercubes can be characterized as bipartite Hamming graphs; see Exercise 2.)

More precisely, a graph G is a Hamming graph if there exist integers $k, n_1, n_2, \ldots, n_k$ such that

$$G \cong K_{n_1} \,\square\, K_{n_2} \,\square\, \cdots \,\square\, K_{n_k}.$$

Letting $V(K_m) = \{1, 2, \ldots, m\}$, the vertex set of G is thus the set of all k-tuples $(i_1, i_2, \ldots, i_k)$, where $i_j \in \{1, 2, \ldots n_j\}$. Two k-tuples are adjacent if and only if they differ in exactly one coordinate.

2.2   Hanoi Graphs

We now consider the celebrated (generalized) *Tower of Hanoi puzzle* [61]. In its general form, the puzzle consists of $p \geq 3$ pegs numbered $1, 2, \ldots, p$ and n discs numbered $1, 2, \ldots, n$, where the discs are ordered by size, disc 1 being the smallest one. A *state*, that is, a distribution of discs on pegs, is called *regular* if on every peg the discs lie in the small-on-large ordering. Figure 2.2 shows an example of a regular state with $p = 4$ pegs and $n = 8$ discs.

Initially, all discs form a *perfect state*—a regular state in that all discs lie on one peg. A *legal move* is a transfer of the topmost disc of a peg to another peg such that no disc is moved onto a smaller one. The objective of the puzzle is to transfer all the discs from one perfect state to another in the minimum number of legal moves.

The natural mathematical object to model the puzzle consists of the Hanoi graphs defined as follows: Let $p \geq 3$, $n \geq 1$, then the *Hanoi graph* H_p^n has all regular states of the p-pegs n-discs problem as vertices, two states being adjacent whenever one is obtained from the other by a legal move. A regular state can be described by the n-tuple $(p_1, p_2, \ldots, p_n)$, where p_i denotes the peg on which the disc i lies. For

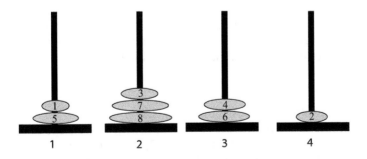

Figure 2.2. A regular state of discs on pegs.

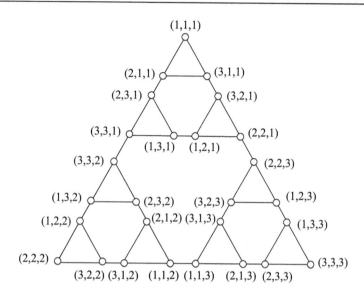

Figure 2.3. The Hanoi graph H_3^3.

instance, the regular state of Figure 2.2 is represented with

$$(1, 4, 2, 3, 1, 3, 2, 2).$$

Conversely, any such n-tuple determines a unique regular state. To see this, we first recall that the coordinates define the disks on each peg. Since the disks on each peg are ordered by their sizes in a regular state, the n-tuple clearly defines a unique regular state. We conclude that there are p^n regular states.

We can consider these n-tuples as vertices of the Hamming graph K_p^n. Since we called two states adjacent when one was obtained from the other by a legal move, this means that adjacent states are represented by n-tuples that differ in exactly one position. In other words, adjacent states correspond to an edge in K_p^n. This means that H_p^n is a spanning subgraph of K_p^n. For instance, moving disc 3 from peg 2 to peg 3 in the regular state of Figure 2.2 yields the state

$$(1, 4, 3, 3, 1, 3, 2, 2).$$

The Hanoi graph H_3^3 is shown in Figure 2.3.

The solution of the puzzle in the case with three pegs is well understood. As can be seen from Figure 2.3, the graphs H_3^n, $n \geq 1$, have a transparent structure. For instance, the figure shows that there is only one shortest path between the vertices $(1, 1, 1)$ and $(3, 3, 3)$ and that its length is $2^3 - 1$. In other words, the solution of the puzzle is unique and requires seven steps. More generally, one can easily guess from here that for n discs, the minimum number of moves is $2^n - 1$. (See Exercise 9.)

In contrast, the puzzle with at least four pegs is a notorious open problem [21, 79] and the structure of the graphs H_p^n, where $p \geq 4$, is not really understood. For example, not even the diameter of these graphs is known (except in small cases that are reachable by a computer). Therefore, the information that the Tower of Hanoi graphs are spanning subgraphs of Hamming graphs brings at least some structure to the graphs H_p^n. It also demonstrates that subgraphs of Hamming graphs possess a rich and complicated structure.

2.3 Exercises

1. Show that Q_3 is a spanning subgraph of $K_{4,4}$.

2. Show that every bipartite Hamming graph is a hypercube.

3. What is the prism over Q_n?

4. Determine the number of induced 4-cycles in Q_n.

5. Show that the distance between two vertices of a hypercube is the number of coordinates in which they differ.

6. Show that the distance between two vertices of a Hamming graph is the number of coordinates in which they differ.

   This result is also an immediate consequence of Lemma 12.2, the so-called Distance Lemma. The corresponding distance function is the *Hamming distance*.

7. Let X be a set with n elements, $n \geq 1$. Let G be the graph whose vertices are all the subsets of X, two vertices A and B being adjacent whenever their symmetric difference contains one element. Show that G is isomorphic to Q_n.

8. Draw the Hanoi graph H_4^2.

9. Prove that the Tower of Hanoi puzzle with 3 pegs and n discs can be solved in $2^n - 1$ moves.

Part II

Classic Topics

3 Hamiltonian Graphs

Many problems in combinatorial optimization are modeled by a cycle containing all the vertices of an appropriate graph. For example, if S is a set of cardinality n, then it could be useful to have a listing $A_1, A_2, \ldots, A_{2^n}$ of all subsets of S such that there is minimal change from any one subset to the next one in the list. In fact, this can be done so that for each index i, the set A_{i+1} (indices modulo 2^n) is obtained from A_i by either inserting or deleting a single element of S. The list so created corresponds in a natural way to a cycle in the n-dimensional cube Q_n. Such a sequence of binary n-tuples is called a *Gray code*.

We first consider several conditions for a graph to have a spanning cycle. In Section 3.2 we turn our attention to the question of when a Cartesian product of two graphs is hamiltonian. It turns out that the existence of a spanning cycle in a Cartesian product is closely related to the spanning trees of the two factors. We then conclude this chapter in Section 3.3 with a brief study of prisms.

3.1 Conditions for Hamiltonicity

A graph G is said to be *hamiltonian*[1] if G has a cycle that contains each of its vertices. We call such a cycle a *hamiltonian cycle*. A graph G is called *traceable* if it has a path containing all of its vertices. Such a path is called a *hamiltonian path*.

An attempt to find a simply stated and useful characterization of hamiltonian and traceable graphs has motivated much research in graph theory during the last century. No nice characterization of the class of hamiltonian or traceable graphs has been found. Many sufficient conditions have been discovered that imply the existence of a

---

[1]The name hamiltonian is a recognition of the Irish mathematician Sir William Rowan Hamilton, who invented a game that required the players to find such a cycle through all the vertices of a dodecahedron by moving along edges of the solid.

hamiltonian cycle or a hamiltonian path. Best known is the following classic result of Dirac [28], where, as usual, $\delta(G)$ denotes the minimum degree of a graph G.

Theorem 3.1. [28] *If a graph G has order $n \geq 3$ and $\delta(G) \geq n/2$, then G is hamiltonian.*

One way to prove this result is to consider a longest path $v_1, v_2, \ldots,$ v_m in the graph. Then use the minimum degree condition to show that v_1 must be adjacent to some v_{j+1} while v_j is a neighbor of v_m. It is then a simple matter to conclude that a cycle can be formed that spans the vertices of this path, and indeed, by the assumption that it is a longest path, we infer $m = n$ (i.e., the cycle spans all of G).

Other than connectivity, the simplest to state and prove—and for our purposes most useful—necessary condition for a graph to be hamiltonian is the following result.

Lemma 3.2. *Let G be a hamiltonian graph and A a proper subset of $V(G)$. Then the graph $G - A$ has at most $|A|$ connected components.*

Proof: For a fixed hamiltonian cycle in G, let C be the spanning subgraph of G that remains when all edges that are not in the hamiltonian cycle are discarded. Let A be a subset of $V(G)$ such that $|A| = k$ for some $1 \leq k < |G|$. Since C is a cycle, if no pair of vertices in A are consecutive on C, then $C - A$ has exactly k components. Otherwise, $C - A$ has fewer than k. The result now follows since $G - A$ has no more components than $C - A$. □

Since the removal of a single vertex from a hamiltonian graph can leave at most one component, it is clear that a hamiltonian graph is 2-connected. However, neither high connectivity nor large minimum degree (larger than some absolute constant) is enough to force a graph to be hamiltonian. For example, $K_{m,m+1}$ is not hamiltonian, but is m-connected and $\delta(K_{m,m+1}) = m$. (See Exercise 1.)

3.2   Hamiltonian Products

The main problem we consider is whether a Cartesian product $G \square H$ is hamiltonian. For this reason, it is natural to limit our discussion to conditions on the factor graphs and how that influences whether their Cartesian product is hamiltonian. For $G \square H$ to be hamiltonian, it is obvious that both G and H must be connected. Hence we assume throughout this chapter that all graphs under consideration are connected.

It is perhaps not surprising that we cannot give a complete characterization of the pairs of graphs whose product is hamiltonian. However, the property of being hamiltonian is preserved when taking Cartesian products. The proof is not difficult and is left to the reader as Exercise 8.

Theorem 3.3. *If G and H are both hamiltonian, then $G \square H$ is also hamiltonian.*

For the remainder of this chapter we consider Cartesian products for which at most one of the factors is hamiltonian. In 1982, Batagelj and Pisanski [8] completely settled the issue when the nonhamiltonian factor is a tree.

Theorem 3.4. [8] *Let T be a tree and H a hamiltonian graph of order n. Then $T \square H$ is hamiltonian if and only if $\Delta(T) \le n$.*

Proof: Let x be a vertex of maximum degree $\Delta = \Delta(T)$ in T with neighbors $t_1, t_2, \ldots, t_\Delta$. We assume without loss of generality that $h_1, h_2, \ldots, h_n, h_1$ is a spanning cycle of H. In what follows we will have occasion to do addition modulo n on the subscripts of $V(H)$.

Every edge e incident with a vertex in the fiber ${}^{t_i}H$ is either an edge in this fiber or is incident with a vertex in the fiber ${}^xH$. Since every $(t_i, h_1), (t_j, h_1)$-path ($i \ne j$) must have an edge incident with a vertex not in the fiber ${}^{t_i}H$, it follows that for $1 \le i < j \le \Delta$, every path in $T \square H$ from (t_i, h_1) to (t_j, h_1) contains at least one vertex from ${}^xH$. Therefore, when ${}^xH$ is deleted from $T \square H$, the resulting graph has at least Δ components. If $\Delta > n$, then $T \square H$ is not hamiltonian by Lemma 3.2.

Now assume that $n \ge \Delta$. Consider any edge-coloring c of the tree T that uses at most n colors. We will assume these n colors are the elements of $\{1, 2, \ldots, n\}$. It is easy to prove by induction that T has such a coloring. (See Exercise 6.) Let V_1, V_2 be the unique bipartition of $V(T)$ into two independent sets (see p. 55). We begin with the vertex set $V(T \square H)$ and construct a hamiltonian cycle by adding edges from $T \square H$ based on a successor function. That is, we add exactly the set of directed edges joining (t, h_j) to $s(t, h_j)$ for all vertices t of T and for all $1 \le j \le n$. This function $s : V(T \square H) \to V(T \square H)$ is a bijection, defined as follows:

Let $1 \le i \le n$, $u \in V_1$, and $v \in V_2$.

- If there is an edge uw in T such that $c(uw) = i$, then we set $s(u, h_i) = (w, h_i)$. If there is no such edge, we define $s(u, h_i) = (u, h_{i+1})$.

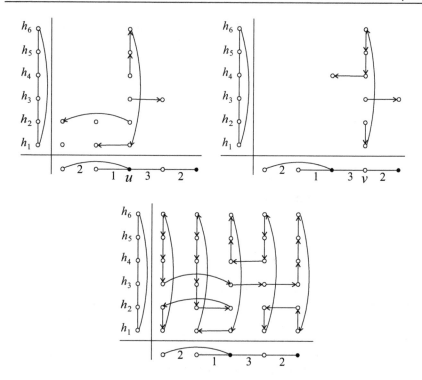

Figure 3.1. An example for Theorem 3.4. See text for explanation.

- If there is an edge vw in T such that $c(vw) = i - 1$, then let $s(v, h_i) = (w, h_i)$. If there is no such edge, we define $s(v, h_i) = (v, h_{i-1})$.

Figure 3.1 illustrates the construction of this directed hamiltonian cycle for a tree of order five and the cycle C_6. The vertices of V_1 are indicated by solid dots and those of V_2 by open circles.

The top left part of Figure 3.1 shows the directed edges arising from the successor of the vertices in the C_6-fiber determined by the vertex $u \in V_1$ according to the first rule.

The top right part of the figure shows the directed edges that are determined according to the second rule by the successor function s applied to vertices in $^vC_6$, where v is one of the vertices in V_2.

The bottom part of the figure shows the entire directed hamiltonian cycle. Note that every edge of T lifts to a pair of oppositely directed edges in two adjacent T-fibers.

Let $x \in V_1$ and let $1 \le j \le n$. The vertex (x, h_j) has a predecessor under s from its T-fiber exactly when x is incident with an edge of color $j - 1$ in T. Since c is an edge-coloring, such an (x, h_j) has only one predecessor in its T-fiber. By the definition of s, we see that in this case, (x, h_j) has no predecessor in its H-fiber. Conversely, if (x, h_j) has a predecessor in its H-fiber, then that predecessor is (x, h_{j-1}) and by definition no edge of color $j - 1$ is incident with x. A similar argument shows that for $y \in V_2$, the vertex (y, h_j) has a unique predecessor under s.

The completion of the proof that the collection of edges constructed as above are the edges of a hamiltonian cycle of $T \square H$ requires a demonstration that the function s is a permutation of $V(T \square H)$ of period $n|T|$. This proof is left to the reader as Exercise 9. $\square$

Note that in the second part of the proof we did not use any edges of H except those in the hamiltonian cycle. In addition, it follows from Theorem 3.4 that for every tree T there exists a "threshold" order $n = \Delta(T)$ such that for every hamiltonian graph H of order at least n, the Cartesian product $T \square H$ is hamiltonian. Conversely, when the order of H is smaller than n, the product is not hamiltonian.

Suppose now that G has a spanning tree T whose maximum degree is at most the order of a hamiltonian graph H. Using only edges of G that belong to T we infer that $T \square H$, and therefore also $G \square H$, is hamiltonian. Hence, the following graphical invariant is natural to study in this context. The *cyclic hamiltonicity*, denoted $cH(G)$, of a connected graph G, is defined as the smallest integer n such that $G \square C_n$ is hamiltonian.

By letting $\nabla(G)$ denote the smallest integer k such that G has a spanning tree whose maximum degree is k, we have the ensuing corollary to Theorem 3.4. The corollary is verified by noting that if $k = \nabla(G) \geq 3$, then G has a spanning tree T with maximum degree k. By Theorem 3.4, we see that $T \square C_k$, and hence also $G \square C_k$, is hamiltonian. The result is that $cH(G) \leq \nabla(G)$.

Corollary 3.5. *If H is a hamiltonian graph and G is a connected graph such that $\nabla(G) \leq |H|$, then $G \square H$ is hamiltonian. In particular, $cH(G) = 3$ if $\nabla(G) = 2$, and $cH(G) \leq \nabla(G)$ if $\nabla(G) \geq 3$.*

A corollary of the following pretty theorem, due to Dimakopoulos, Palios and Poulakidas [27], shows that $\nabla(G)$ cannot be larger that $cH(G) + 1$.

Theorem 3.6. [27] *If G and H are connected graphs such that $G \square H$ is hamiltonian, then $\nabla(G) \leq |H| + 1$ and $\nabla(H) \leq |G| + 1$.*

Proof: We will use a spanning cycle C of $G \square H$ to construct a spanning tree T of G such that $\Delta(T) \leq |H| + 1$. As we traverse C, we will maintain

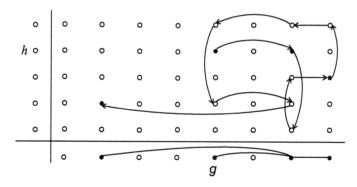

Figure 3.2. Vertices of G in M and edges of F after a partial traversing of the spanning cycle of $G \,\square\, H$.

a set M of "marked" vertices of G and a subset F of the edge set of G. When all the vertices of $G \,\square\, H$ have been visited on C, M will be $V(G)$ and F will be the edges of the spanning tree T. Initially, both of M and F are the empty set. Select an arbitrary vertex (g, h), and add g to M. Move along C, and for every pair (g', h') and (g'', h'') of consecutive vertices encountered on C, perform the following:

- If g'' is not marked (i.e., $g'' \notin M$), then add g'' to M and add the edge $g'g''$ to F.

The construction is illustrated in Figure 3.2.

Note in the above, that if g'' was not marked when vertex (g'', h'') was reached from (g', h') along C, then it follows from the edge structure of the Cartesian product that $g'g''$ is an edge of G and $h' = h''$. By the way edges are added to F, it is clear that when C has been completely traversed, $M = V(G)$, $|F| = |G| - 1$, and the subgraph T induced by F has no cycles. Hence, T is a spanning tree of G. The vertex g is incident with at most $|H|$ of the edges of T. These edges were added to F when C moved from the fiber ${}^g H$ to a different fiber ${}^x H$ when x was not marked. Other vertices u of G can be incident with at most $|H| + 1$ edges of T, one when ${}^u H$ was first entered by C and at most $|H|$ additional edges added when C left ${}^u H$ for another H-fiber that had not yet been entered by C.

Therefore, we have constructed a spanning tree T of G and

$$\nabla(G) \le \Delta(T) \le |H| + 1.$$

Similarly, $\nabla(H) \le |G| + 1$. $\square$

Corollary 3.7. *Let G be a connected graph. Then $\nabla(G) \le cH(G) + 1$.*

Proof: Let $\nabla(G) = k$. Then $\nabla(G) > k - 1 = |C_{k-2}| + 1$. By Theorem 3.6, $G \,\square\, C_{k-2}$ is not hamiltonian. Thus, $\mathrm{cH}(G) \geq k - 1 = \nabla(G) - 1$. □

The complete bipartite graph $G = K_{2,2m}$ shows that the upper bound in Corollary 3.7 is tight. (See Exercise 11.)

It is not necessary that either factor be hamiltonian for their Cartesian product to be hamiltonian. In fact, the smallest nontrivial Cartesian product, the cycle of order four, already demonstrates this.

3.3 Hamiltonian Prisms

We recall that the prism over a graph G is the Cartesian product $G \,\square\, K_2$. As the structure of $G \,\square\, K_2$ is quite straightforward, we would think that it should not be too difficult to decide whether the prism over G is hamiltonian. Note that if G is traceable, then the prism over G is hamiltonian. (See Exercise 7.) However, the prism over G can still be hamiltonian even if G is not traceable. (See Exercise 12.) This indicates that the problem might not be easy.

Let us restrict our attention to planar graphs now. By a famous theorem of Tutte [109], every 4-connected planar graph is hamiltonian. Hence the prism over such a graph is hamiltonian as well. In contrast, there exist 2-connected planar graphs whose prisms are not hamiltonian. (See Exercise 13.) For the remaining case of 3-connected planar graphs, Rosenfeld and Barnette [102] advanced the following conjecture in 1973.

Conjecture 3.8. [102] *Let G be a 3-connected planar graph. Then the prism over G is hamiltonian.*

The conjecture is still open, but recently Biebighauser and Ellingham [11] succeeded in proving the conjecture for 3-connected bipartite planar graphs and for triangulations of the plane, the projective plane, the torus, and the Klein bottle.

For a graph G, let us call the Cartesian product $G \,\square\, Q_k$, $k \geq 1$, the k-tuple prism over G. Note that the 1-tuple prism over G is just the prism over G. So the Rosenfeld-Barnette conjecture asserts that the 1-tuple prism over a 3-connected planar graph is hamiltonian. Even though this seems to be quite a difficult problem, we can (a bit surprisingly) prove that for $k \geq 2$, the k-tuple prism over a 3-connected planar graph is hamiltonian. To prove it, we will invoke the following theorem of Barnette [7].

Theorem 3.9. [7] *Every 3-connected planar graph has a spanning tree T with $\Delta(T) \leq 3$.*

Now we can state the announced theorem. Instead of following the original proof, we make use of Theorem 3.4.

Theorem 3.10. [102] *Let G be a 3-connected planar graph, and let $k \geq 2$. Then the k-tuple prism over G is hamiltonian.*

Proof: By Theorem 3.9, G contains a spanning tree T with $\Delta(T) \leq 3$. Clearly, as G is 3-connected, it contains at least 4 vertices and $\Delta(T) \geq 2$. Q_k, $k \geq 2$, is hamiltonian and has order $2^k \geq 3 \geq \Delta(T)$. Hence, Theorem 3.4 implies that $T \square Q_k$ is hamiltonian. As $T \square Q_k$ is a spanning subgraph of $G \square Q_k$, the latter graph is hamiltonian as well. $\square$

3.4   Exercises

1. Show that a bipartite graph of odd order is not hamiltonian.

2. Show that the grid graph $P_n \square P_m$ is hamiltonian if and only if at least one of n or m is even.

3. Let us say that a bipartite graph with color classes C_1 and C_2 is *color-balanced* if $|C_1| = |C_2|$. Let G and H be connected bipartite graphs such that $G \square H$ is hamiltonian. Show that at least one of G or H is color-balanced.

4. Show how to construct a spanning cycle of $K_{1,n} \square C_{2n}$ for $n \geq 2$. (This shows that one of the bipartite factor graphs can be far from color-balanced when their Cartesian product is hamiltonian; see Exercise 3.)

5. [9] Assume that each of G and H has a vertex adjacent to two or more vertices of degree one. Show that $G \square H$ is not hamiltonian.

6. Let T be a tree having maximum degree k. Prove that there is an assignment of colors from the set $\{1, 2, \ldots, k\}$ to the edges of T such that adjacent edges are given distinct colors.

7. Show that if G and H are both traceable, then $G \square H$ is hamiltonian unless both are bipartite of odd order.

8. Prove Theorem 3.3. (A good way to begin is to consider two cases based on the parities of the factors.)

9. Complete the proof of Theorem 3.4 by showing that the function s, when viewed as a permutation of $V(T \square H)$, consists of a single cycle having period $n|T|$.

10. [9] Show that if G has three vertices of degree one that have a common neighbor and H has at least one vertex of degree one, then $G \mathbin{\square} H$ is not hamiltonian.

11. Show that $\nabla(K_{2,2m}) = m + 1$ and construct a hamiltonian cycle for $K_{2,2m} \mathbin{\square} C_m$, thus showing the bound in Corollary 3.7 is tight.

12. Show that the prism over $K_{2,4}$ is hamiltonian.

13. Show that the prism over $K_{2,5}$ is not hamiltonian.

4

Planarity and Crossing Number

In this chapter, we consider the question of how well Cartesian product graphs fit into the plane. We first characterize planar Cartesian products; not too surprisingly, there are only a few. Standard examples of such graphs are the grid graphs $P_m \square P_n$. Conversely, products of cycles are not planar, so it is reasonable to ask what is the smallest number of crossings in their plane drawings. In Section 4.2, we thus consider the famous conjecture asserting that an optimal plane drawing of $C_m \square C_n$, $n \geq m \geq 3$, requires $(m-2)n$ crossings. In the concluding section, we list several results, mostly recent, that give exact crossing numbers of special Cartesian products.

When considering planar Cartesian products and crossing numbers, we may clearly restrict ourselves to connected graphs; see Exercise 1. Hence, all graphs in this chapter are connected.

4.1 Planar Products

In this section we show that the only planar Cartesian products are $P_m \square P_n$, $P_m \square C_n$, and $G \square K_2$, where G is outerplanar. This result was proved in 1969 by Behzad and Mahmoodian [9]. We first consider factors on at least three vertices.

Proposition 4.1. [9] *Let G and H be connected graphs on at least three vertices. Then $G \square H$ is planar if and only if both G and H are paths or if one is a path and the other a cycle.*

Proof: Consider first the cases where the vertices of both factors have degree at most two, that is, when both factors are paths or cycles. Clearly, $P_m \square P_n$ is planar for any $m, n \geq 3$. It is also easy to find planar drawings of $P_m \square C_n$, $m, n \geq 3$. (For instance, if we remove the seven bent edges of Figure 4.3 later in this chapter, we obtain a plane

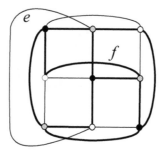

Figure 4.1. A subdivision of $K_{3,3}$ as a subgraph of $C_3 \square C_3$.

drawing of $P_4 \square C_7$.) In Figure 4.1, we can see a subdivision of $K_{3,3}$ as a subgraph of $C_3 \square C_3$; hence, $C_3 \square C_3$ is not planar. Similarly, for $m, n \geq 3$, $C_m \square C_n$ contains a subdivision of $K_{3,3}$ and so is not planar.

Suppose now that at least one of the factors has a vertex of degree at least three. Then $G \square H$ contains $K_{1,3} \square P_3$ as a subgraph, which in turn contains a subdivision of $K_{3,3}$ (see Exercise 2), so no such product is planar.   □

There is nothing to be done if one factor has a single vertex since $K_1 \square G \cong G$ for any G. Hence, we are left with the case when one factor is K_2. To be able to consider this case, recall that a graph G is *outerplanar* if it can be drawn in the plane in such a way that every vertex lies on the boundary of the unbounded face. Using Kuratowski's theorem, Chartrand and Harary [20] showed that a graph is outerplanar if and only if it has no subgraph that is a subdivision of K_4 or $K_{2,3}$. For the next result, the following lemma (see Exercise 4) will be useful.

Lemma 4.2. *Let u be a vertex of a graph H, and let X be a graph constructed from H and a new vertex x that is adjacent to u and all its neighbors. If a graph G contains a subdivision of H, then $G \square K_2$ contains a subdivision of X.*

Proposition 4.3. [9] *Let G be a connected graph. Then $G \square K_2$ is planar if and only if G is outerplanar.*

Proof: Suppose that G is outerplanar. Then $G \square K_2$ can be drawn in the plane as indicated in Figure 4.2.

Conversely, let $G \square K_2$ be planar. If G contains a subdivision of K_4 or a subdivision of $K_{2,3}$, then by Lemma 4.2, $G \square K_2$ would contain a subdivision of K_5 or a subdivision of $K_{3,3}$. It follows that G is outerplanar.   □

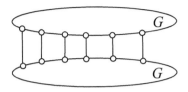

Figure 4.2. $G \square K_2$ is planar provided G is outerplanar.

Thus, only some Cartesian products are planar. However, we would also like to draw nonplanar products as nicely as possible in the plane. The first obvious case to consider is the Cartesian product of two cycles.

4.2   Crossing Numbers of Products of Cycles

The *crossing number*, cr(G), of the graph G is the minimum number of crossings over all drawings of G. An optimal drawing of G (that is, a drawing with the minimum number of crossings) always has the following properties:

(i)   no two adjacent edges intersect,

(ii)   no edge intersects itself,

(iii) no two edges intersect each other more than once, and

(iv)   each intersection of edges is a crossing (that is, it is not simply a tangency).

Moreover, every optimal drawing of G can be modified in such a way that no three edges intersect in a common point.

For instance, the drawing of $C_3 \square C_3$ of Figure 4.1 has three crossings and hence cr($C_3 \square C_3$) ≤ 3. We will see later that no better drawing of $C_3 \square C_3$ exists. Before we do that, however, let us consider the drawing of $C_4 \square C_7$ from Figure 4.3. This drawing has $2 \cdot 7$ crossings and can easily be extended to a drawing of $C_m \square C_n$ ($n \geq m \geq 3$) with $(m-2)n$ crossings.

Hence, cr($C_m \square C_n$) $\leq (m-2)n$ for $n \geq m \geq 3$. It has been conjectured by Harary, Kainen, and Schwenk [52] that this is the best possible bound.

Conjecture 4.4. [52] *For any* $n \geq m \geq 3$, cr($C_m \square C_n$) $= (m-2)n$.

In the rest of this section we prove the conjecture for $m = 3$ and all n. For the sake of this proof, we first show that the drawing of Figure 4.1 is optimal.

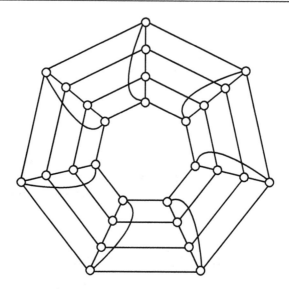

Figure 4.3. A drawing of $C_4 \square C_7$.

Lemma 4.5. [52] *The crossing number of $C_3 \square C_3$ is 3.*

Proof: To shorten the notation, set $G = C_3 \square C_3$. We already know that G is not planar. Moreover, if we could draw G with only one crossing, then we could remove one edge of G to obtain a planar graph. But in Figure 4.1, we see that there exists an edge that can be removed without destroying the subdivision of $K_{3,3}$. (Note that by the symmetry of G, this can be any edge.) Hence, $2 \leq \text{cr}(C_3 \square C_3) \leq 3$.

Suppose $\text{cr}(C_3 \square C_3) = 2$. We select an arbitrary optimal drawing with two crossings. Assume that e and f are edges of this drawing that cross, where e and f project on the same factor, say the first one. By property (i) of optimal drawings, e and f lie in different triangles, say T_e and T_f, which both project onto the first factor. In completing the drawing of T_e and T_f, another crossing occurs; see Figure 4.4.

Moreover, two vertices of one of the triangles are separated by the other one. See Figure 4.4, in which the triangle $T_e = xyz$ with $e = xz$

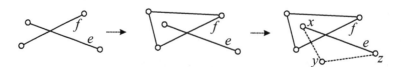

Figure 4.4. A drawing of two crossings.

is drawn such that x is inside the other triangle while y and z are outside. Note that in $C_3 \square C_3$ there exists an x, z-path P that is internally disjoint with T_e and T_f. But then P requires another crossing. Hence, if $\mathrm{cr}(C_3 \square C_3) = 2$, then two edges that cross necessarily project on different factors.

So in any crossing, one edge projects on the first factor and the other edge on the second factor. Let e be the edge of the first crossing that projects onto the first factor and f the edge of the second crossing that projects onto the second factor. Since we have assumed that we have a drawing with two crossings, $G - \{e, f\}$ is planar. Now, e and f cannot be adjacent. See Figure 4.1 for two such edges whose removal leaves the subdivision of $K_{3,3}$. (We again use the symmetry of G to argue this for any two adjacent edges that project on different factors.) Similarly, e and f cannot be at distance one. So e and f must be at distance two, and this is indeed possible; see Figure 4.1, where the removal of e and f yields a planar graph. (Once again, by symmetry, there is only one way to select e and f.) The obtained graph is 3-connected (see Chapter 5 for the definition); hence, by Whitney's theorem,[1] its embedding in the plane is unique. But then we need at least three crossings in order to add the edges e and f back to the drawing of $G - \{e, f\}$.   □

Before presenting the next theorem, one more lemma is needed. We set $V(C_n) = \{1, 2, \ldots, n\}$.

Lemma 4.6. *Let $n \geq 4$ and suppose that in a drawing of $C_3 \square C_n$ no fiber $(C_3)^i$, $1 \leq i \leq n$, has a crossed edge. Then the drawing contains at least n crossings.*

Proof: Let $G = C_3 \square C_n$ and let G_i be the subgraph of G induced by the fibers $(C_3)^i$ and $(C_3)^{i+1}$, $1 \leq i \leq n$. In particular, G_n is induced by $(C_3)^n$ and $(C_3)^1$. We claim that in G_i there are at least two edges that are crossed. More precisely, if there are two edges of G_i that cross, we have one crossing; otherwise, at least two edges of G_i (multiple occurrences of the same edge are allowed) each cross with some edge not in G_i. Since there are n subgraphs G_i, the proof will be complete after establishing the claim.

If two of the edges of G_i cross, there is nothing to be proved. So assume no two edges of G_i cross. Then, by Whitney's theorem, $G_i = C_3 \square K_2$ is uniquely embeddable into the plane because it is 3-connected. It follows that the fiber $(C_3)^{i+2}$ lies entirely in one of the

---

[1]Whitney's theorem asserts that a 3-connected planar graph has only one planar representation in the sense that the facial cycles are uniquely determined [93, Theorem 2.5.1].

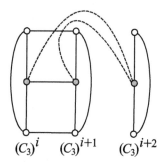

Figure 4.5. Two edges of G_i are crossed.

quadrangular faces of G_i, say in the outside face F. (Note that the fiber $(C_3)^{i+2}$ cannot lie in a triangular face because, by the assumption, no fiber $(C_3)^i$, $1 \le i \le n$, has a crossed edge.) See Figure 4.5, where the situation is presented.

Consider the C_n-fiber that contains the three gray vertices of Figure 4.5. Then this cycle must cross the boundary of F at least twice. Hence at least two edges of G_i are crossed. (It is again possible that the same edge is counted twice.)   □

We are now in a position to prove the announced theorem of Ringeisen and Beineke [101].

Theorem 4.7. [101] *For any $n \ge 3$, $\mathrm{cr}(C_3 \,\square\, C_n) = n$.*

Proof: The assertion for $n = 3$ is Lemma 4.5. Suppose the result holds for $k \ge 3$, and consider a drawing of $C_3 \,\square\, C_{k+1}$. Assume the drawing has fewer than $k + 1$ crossings. By Lemma 4.6, this is possible only if there exists a fiber $(C_3)^i$ with a crossed edge. Let H be the graph obtained from $C_3 \,\square\, C_{k+1}$ by deleting the edges of the fiber $(C_3)^i$. But the drawing of H yields a drawing of $C_3 \,\square\, C_k$ with fewer than k crossings, in contradiction to the induction hypothesis.   □

Note that the induction basis for Theorem 4.7, that is, Lemma 4.5, is as demanding as the induction step. The proof idea of Theorem 4.7 can be applied to verify Conjecture 4.4 also for bigger n. However, the details are more involved. For example, the basis of the induction for $n = 5$ ($\mathrm{cr}(C_5 \,\square\, C_5) = 15$) is proved by Richter and Thomassen [100], and the induction step is then given by Klešč, Richter, and Stobert [84]. Recently, Adamsson and Richter [1] proved the conjecture for $m = 7$. Interestingly, their approach can also be used for the induction step for the case $m = 8$, but the basis ($\mathrm{cr}(C_8 \,\square\, C_8) = 48$?) is still open. Glebsky and Salazar [43] used the approach of Adamsson and Richter [1] to obtain the following breakthrough on the conjecture.

Theorem 4.8. [43] *For any $m \geq 3$ and $n \geq m(m+1)$,*

$$\operatorname{cr}(C_m \square C_n) = (m-2)n.$$

This result thus asserts that, for any $m \geq 3$, Conjecture 4.4 holds for all but at most finitely many n.

4.3   More Exact Crossing Numbers

The crossing number has also been determined for several additional Cartesian products. We next list three such results as theorems. The proofs for most of these results are rather technical and are thus omitted. In fact, each of the presented results requires a separate paper. Usually, but not necessarily, the easier part is to find optimal drawings.

The first result, due to Klešč [82], seems rather specialized but is far from being trivial.

Theorem 4.9. [82] *The crossing number of $K_{2,3} \square C_3$ is 9.*

The following result (Theorem 4.10) was conjectured in 1982 by Jendrol and Ščerbová [72] and proved recently by Bokal [12].

Theorem 4.10. [12] *For any $m, n \geq 1$,*

$$\operatorname{cr}(K_{1,m} \square P_n) = (n-2) \left\lfloor \frac{m}{2} \right\rfloor \left\lfloor \frac{m-1}{2} \right\rfloor.$$

In fact, using a similar approach, Bokal extended Theorem 4.10 to Cartesian products of stars and trees [13]. He obtained an expression that gives explicit crossing numbers for products of stars with trees that have a maximum degree of at most four, as well as for the products $K_{1,3} \square T$ and $K_{1,4} \square T$, where T is an arbitrary tree.

Recently, Peng and Yiew [98] proved a result (Theorem 4.11) that deals with Cartesian products of three factors.

Theorem 4.11. [98] *For any $n \geq 1$,*

$$\operatorname{cr}(P_n \square K_2 \square K_3) = 4(n-1).$$

Several additional exact crossing numbers can be found in the literature. Here we mention only the paper by Klešč [83], in which the crossing numbers of the Cartesian products of four special graphs of order five with cycles are given. An interested reader can also find a survey of related results in the same paper.

We conclude this chapter by pointing out that $C_m \square C_n$ can be naturally drawn on the torus without crossings. For this reason, Cartesian products of cycles are also known as the *torus graphs*. Since they can be drawn on the torus we also say that their genus is 1, but further discussion is beyond the scope of the book.

4.4   Exercises

1. Let G and H be arbitrary graphs, not necessarily connected. Describe the connected components of $G \square H$.

2. Find a subdivision of $K_{3,3}$ in $K_{1,3} \square P_3$.

3. [74] Show that the Cartesian product of two connected graphs is outerplanar if and only if one factor is a path and the other is K_2.

4. Prove Lemma 4.2.

5. Determine $\mathrm{cr}(K_5)$ and $\mathrm{cr}(K_{3,3})$.

6. Find a drawing of $K_{2,3} \square C_3$ with 12 crossings.

7. For any $m, n \geq 1$, find a drawing of $(K_{1,m} \square P_n)$ with

$$(n - 2) \left\lfloor \frac{m}{2} \right\rfloor \left\lfloor \frac{m-1}{2} \right\rfloor$$

crossings, as Theorem 4.10 asserts.

8. For any $n \geq 1$, find a drawing of $P_n \square K_2 \square K_3$ with $4(n-1)$ crossings, as Theorem 4.11 asserts.

5 Connectivity

Connectivity is one of the fundamental and central concepts in graph theory. It is important from both a theoretical and a practical point of view. For instance, in order for a communication network or a distributed process to be robust, the underlying graph must have large enough connectivity. Since hypercubes and other products are frequently used models of communication networks, we are thus led to the investigation of the connectivity of Cartesian products.

A slightly more detailed view of this stability problem shows that in a communications network, at least two matters can go wrong. First, elements of the network that are represented by vertices can be damaged. In this case, the (vertex) connectivity of the network is essential. Second, if there are problems with communication channels, then the edge-connectivity of the network should be large enough to guarantee network stability.

In this chapter, we show that the connectivity and the edge-connectivity of a Cartesian product depend only on the connectivities, the minimum degrees, and the orders of the factors.

5.1 Vertex-Connectivity

Recall that a Cartesian product is connected if and only if all factors are connected; refer to Lemma 1.1. However, as we already mentioned, it is important to know not only whether a graph is connected, but also how strongly it is connected. To investigate this in more detail we recall several concepts.

Let G be a graph. Then $S \subseteq V(G)$ is a *separating set* if $G - S$ has more than one component. The *connectivity* $\kappa(G)$ of G is the minimum size of $S \subseteq V(G)$ such that $G - S$ is disconnected or a single vertex.[1] For any $k \leq \kappa(G)$, we say that G is k-connected.

---

[1] Note that by this definition $\kappa(K_1) = 0$, although K_1 is connected.

For instance, $\kappa(Q_3) = 3$, and hence Q_3 is 1-connected, 2-connected, as well as 3-connected. Note that, with the exception of K_1, every connected graph is 1-connected.

Let u be a vertex of a graph G. If $\deg(u) < |G| - 1$, then the set of neighbors of u forms a separating set. Hence $\kappa(G) \leq \deg(u)$. Moreover, if $\deg(u) = |G| - 1$, then $\kappa(G) \leq |G| - 1 = \deg(u)$. It follows that $\kappa(G)$ is bounded by the minimum degree $\delta(G)$ of G. That is, $\kappa(G) \leq \delta(G)$.

Let S be a separating set of a graph G, and let H be any graph. Then $S \times V(H)$ is a separating set of $G \,\square\, H$ (Exercise 1). Consequently,

$$\kappa(G \,\square\, H) \leq \kappa(G)|H|, \tag{5.1}$$

and, analogously,

$$\kappa(G \,\square\, H) \leq \kappa(H)|G|. \tag{5.2}$$

A graph contains a separating set if and only if it is not a complete graph. Hence (5.1) and (5.2) hold for the case when neither G nor H is complete. Now, let G be an arbitrary graph on at least two vertices and $n \geq 2$. Then

$$\begin{aligned}
\delta(G \,\square\, K_n) &= \delta(G) + n - 1 \leq |G| - 1 + n - 1 \\
&\leq |G|(n - 1) = |G|\kappa(K_n), \tag{5.3}
\end{aligned}$$

where the last inequality holds because it is equivalent to $n - 2 \leq |G|(n - 2)$ for $n \geq 2$. Since

$$\kappa(G \,\square\, H) \leq \delta(G \,\square\, H) = \delta(G) + \delta(H), \tag{5.4}$$

the inequalities (5.1), (5.2), (5.3), and (5.4) combine into

$$\kappa(G \,\square\, H) \leq \min\{\kappa(G)|H|, \kappa(H)|G|, \delta(G) + \delta(H)\} \tag{5.5}$$

for any graphs G and H on at least two vertices.

The minimum in (5.5) can be realized by any of the three terms. (See Exercise 2.) It turns out that this simple upper bound for the connectivity of the Cartesian product is also a lower bound. This is not so obvious and is stated in the following, main result of this chapter.

Theorem 5.1. [88, 107][2] *Let G and H be graphs on at least two vertices. Then*

$$\kappa(G \,\square\, H) = \min\{\kappa(G)|H|, \kappa(H)|G|, \delta(G) + \delta(H)\}.$$

---

[2]This theorem was first stated by Liouville [88]. However, the announced proof never appeared. In the meantime, several partial results were obtained until, finally, Špacapan [107] provided the proof that we follow here.

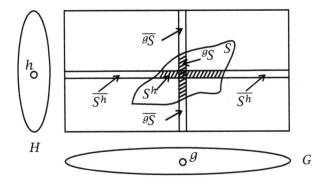

Figure 5.1. Notations for the proof of Theorem 5.1.

Proof: If $\kappa(G) = 0$ or $\kappa(H) = 0$, then $\kappa(G \square H) = 0$ by Lemma 1.1; thus, the result holds if one of the factors is not connected.

Let both G and H be connected graphs. By (5.5), we only need to prove that $\kappa(G \square H)$ is at least the claimed minimum. For this, it suffices to prove that if S is a separating set of $G \square H$ with $|S| < \min\{\kappa(G)|H|, \kappa(H)|G|\}$, then $|S| \geq \delta(G \square H)$. (Note that $G \square H$ is not a complete graph; hence, it contains separating sets.)

Suppose first that for some $g \in G$ and for some $h \in H$, all the vertices of the fibers G^h and ^gH lie in S. Then

$$|S| \geq |G| + |H| - 1 \geq \delta(G) + 1 + \delta(H) + 1 - 1 > \delta(G \square H).$$

In the rest of the proof we may thus, without loss of generality, assume that no G-fiber is induced by a subset of S. Let us introduce the following notations. For $g \in G$, let ^gS be the subgraph of ^gH induced by $V(^gH) \cap S$ and let $\overline{^gS} = {}^gH - {}^gS$. For $h \in H$, the subgraphs S^h and $\overline{S^h}$ are defined analogously. See Figure 5.1.

In these notations, $\overline{S^h} \neq \emptyset$ for every $h \in H$. Since $|S| < \kappa(G)|H|$, there exists an $h \in H$ such that $|S^h| < \kappa(G)$. Therefore, $\overline{S^h}$ is connected. Let C be the connected component of $(G \square H) - S$ that contains $\overline{S^h}$; see Figure 5.2.

Set $X = \{x \in H \mid \overline{S^x} \subseteq C\}$. Since $h \in X$ we have $X \neq \emptyset$. Also, $X \neq V(H)$ because otherwise $(G \square H) - S$ would be connected. Since H is connected, there exist adjacent vertices h'' and h' of H such that h'' is in X but not h'. Without loss of generality, we may assume that $h = h''$; that is, select h in the first place such that it has a neighbor not in X.

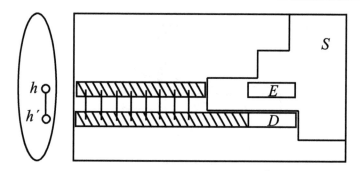

Figure 5.2. Adjacent G-fibers used in the proof of Theorem 5.1.

By the definition of X, $\overline{S^{h'}} \cap C \neq \overline{S^{h'}}$. Let $D = \overline{S^{h'}} - C$; then D is nonempty. See Figure 5.2 again. Since no vertex of D is adjacent to a vertex of C, the neighbor (g,h) of a vertex $(g,h') \in D$ must lie in S^h. Therefore, $|S^h| \geq |D|$. Let E be the isomorphic copy of D in G^h. Then, as just noted, E is contained in S^h, and consequently $|E| \leq |S^h|$. Let x be a vertex from D. Since x is adjacent to no vertex from C, we have

$$\delta(G) \leq \deg_{G^{h'}}(x) \leq |D| - 1 + |S^{h'}| = |E| - 1 + |S^{h'}|,$$

and therefore

$$|S^{h'}| + |E| \geq \delta(G) + 1. \tag{5.6}$$

Suppose that there exists a $g \in G$ such that $^g S = {}^g H$. Since D and $S^{h'}$ are disjoint, $p_G(D)$ and $p_G(S^{h'})$ are disjoint as well. It follows that $^g H$ intersects $S^{h'} \cup E$ in at most one vertex. Consequently, using (5.6), we get

$$|S| \geq |S^{h'}| + |E| + |{}^g H| - 1 \geq \delta(G) + 1 + \delta(H).$$

Assume, finally, that no H-fiber is induced by a subset of S. Then, by arguments analogous to those above, we find two adjacent vertices g and g' of G such that

$$|{}^{g'} S| + |E'| \geq \delta(H) + 1, \tag{5.7}$$

where E' is a subgraph of $^g H$ (included in $^g S$) analogous to the subgraph E of G^h. Since $S^{h'} \cup E$ intersects $^{g'} S \cup E'$ in at most two vertices, (5.6) and (5.7) imply that

$$|S| \geq |S^{h'}| + |E| + |{}^{g'} S| + |E'| - 2 \geq \delta(G) + \delta(H),$$

which completes the proof. $\square$

Theorem 5.1 in particular implies the following result of Sabidussi [103]:

Corollary 5.2. [103] *Let G and H be connected graphs. Then*

$$\kappa(G \,\square\, H) \geq \kappa(G) + \kappa(H).$$

Proof: Since $\kappa(K_1) = 0$, the result holds if G or H is K_1. Suppose G and H are connected graphs on at least two vertices. As the connectivity of a graph is at most its minimum degree, we infer $\delta(G) + \delta(H) \geq \kappa(G) + \kappa(H)$. Since H has at least two vertices,

$$\kappa(G)|H| = \kappa(G) + \kappa(G)(|H| - 1) \geq \kappa(G) + \kappa(H).$$

Analogously, we find that $\kappa(H)|G| \geq \kappa(G) + \kappa(H)$. The result now follows from Theorem 5.1. □

In most cases, the minimum in Theorem 5.1 will be attained by $\delta(G) + \delta(H)$. This holds in particular for all powers of connected graphs, as the next theorem asserts.

Theorem 5.3. *Let G be a connected graph on at least two vertices. Then for any $n \geq 2$,*

$$\kappa(G^n) = n\,\delta(G).$$

Proof: Assume first that $\kappa(G) = 1$, that is, either $G = K_2$ or G contains a cut vertex x. In the latter case, consider a smallest component of $G - x$ to see that $\delta(G) < |G|/2$. Hence, in any case $\delta(G) \leq |G|/2$. Using Theorem 5.1, it follows that $2\delta(G) \leq |G| = \kappa(G)|G|$; thus, $\kappa(G^2) = 2\delta(G)$. Let $n \geq 3$. Then, by induction,

$$\kappa(G \,\square\, G^{n-1}) = \min\{|G|^{n-1}, (n-1)\delta(G)|G|, n\delta(G)\}. \qquad (5.8)$$

Clearly, $n\delta(G) \leq (n-1)\delta(G)|G|$. Moreover, if $G = K_2$, then $n\delta(G) = n \leq 2^{n-1} = |G^{n-1}|$. And if $|G| \geq 3$, then $n\delta(G) \leq n|G| \leq |G^{n-1}| = |G|^{n-1}$ holds since $n \geq 3$ and $|G| \geq 3$. In any case, the minimum in (5.8) equals $n\delta(G)$, so the result holds for powers of graphs with connectivity 1.

Let $\kappa(G) \geq 2$. Then $2\delta(G) \leq \kappa(G)|G|$. Hence, the assertion is true for $n = 2$. Let $n \geq 3$. Then by the induction hypothesis,

$$\kappa(G \,\square\, G^{n-1}) = \min\{\kappa(G)|G|^{n-1}, (n-1)\delta(G)|G|, n\delta(G)\}. \qquad (5.9)$$

Clearly, $n\delta(G) \leq (n-1)\delta(G)|G|$. Moreover,

$$n\delta(G) \leq n(|G| - 1) \leq n|G| \leq 2|G|^{n-1} \leq \kappa(G)|G|^{n-1},$$

where $n|G| \leq 2|G|^{n-1}$ (that is, $n/2 \leq |G|^{n-2}$) holds, since $n \geq 3$ and $|G| \geq 2$. We conclude that the minimum in (5.9) equals $n\delta(G)$. □

Recall that the connectivity of hypercubes was the initial motivation for this chapter.

Corollary 5.4. *For any $n \geq 1$, $\kappa(Q_n) = n$.*

Proof: The assertion is clear for $n = 1$. For $n \geq 2$, recall that Q_n is the Cartesian product of n copies of K_2, and use Theorem 5.3.   □

5.2   Edge-Connectivity

Let G be a graph. Then $S \subseteq E(G)$ is a *disconnecting set* if $G - S$ has more than one connected component. Note that if G is not connected, then the empty set of edges is disconnecting. G is *k-edge-connected* if every disconnecting set consists of at least k edges, while the *edge-connectivity* of G, $\kappa'(G)$, is the maximum k for which G is k-edge-connected. If G is not connected, then by the above remark, $\kappa'(G) = 0$.

Since for any graph G, $\kappa'(G) \geq \kappa(G)$ (Exercise 7), Theorem 5.1 implies that for any graphs G and H on at least two vertices,

$$\kappa'(G \square H) \geq \min\{\kappa(G)|H|, \kappa(H)|G|, \delta(G) + \delta(H)\}. \tag{5.10}$$

The main insight of this section asserts that κ can be replaced with κ' in (5.10).

Clearly, for any graph G, $\kappa'(G) \leq \delta(G)$, and hence

$$\kappa'(G \square H) \leq \delta(G \square H) = \delta(G) + \delta(H). \tag{5.11}$$

Let S be a disconnecting set of edges of a graph G and let H be any graph. Then the set of edges of $G \square H$ that project on S form a disconnecting set in $G \square H$ (Exercise 9). Consequently,

$$\kappa'(G \square H) \leq \kappa'(G)|H| \tag{5.12}$$

and, analogously,

$$\kappa'(G \square H) \leq \kappa'(H)|G|. \tag{5.13}$$

Inequalities (5.11), (5.12), and (5.13) prove the easier part of the main result of this section. It was proved by Xu and Yang [119]. Here we follow the short, elegant argument of Špacapan [106].

Theorem 5.5. [119] *Let G and H be graphs on at least two vertices. Then*

$$\kappa'(G \square H) = \min\{\kappa'(G)|H|, \kappa'(H)|G|, \delta(G) + \delta(H)\}.$$

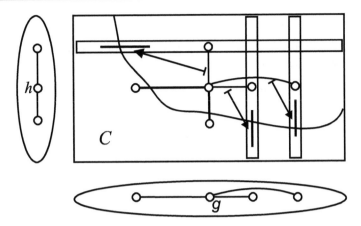

Figure 5.3. Assigning edges from S to the neighbors of (g, h).

Proof: Note that the result is clearly true if G or H is not connected. So assume in the rest of the proof that G and H are connected graphs on at least two vertices.

Let $S \subseteq E(G \square H)$ be an arbitrary disconnecting set of $G \square H$ with $|S| = \kappa'(G \square H)$. If $|S| \geq \min\{\kappa'(G)|H|, \kappa'(H)|G|\}$ the proof is done. Hence, assume that $|S| < \min\{\kappa'(G)|H|, \kappa'(H)|G|\}$. We need to prove that $|S| \geq \delta(G \square H)$.

Since $|S| < \kappa'(G)|H|$, there exists a G-fiber that is connected in $(G \square H) - S$. Analogously, there is an H-fiber that is connected in $(G \square H) - S$. Let C be the connected component of $(G \square H) - S$ that contains these two fibers (note that the two fibers are in the same connected component) and let (g, h) be an arbitrary vertex of $G \square H$ that is not in C. (Such a vertex exists since S is a disconnecting set.)

We claim that $\deg(g, h) \leq |S|$ and prove this fact by assigning to each neighbor of (g, h) a unique edge from S. So let (g', h) be a neighbor of (g, h) in G^h. If $e = [(g, h), (g', h)] \in S$ we assign e to (g', h). Suppose $[(g, h), (g', h)] \notin S$. Then $(g', h) \notin C$, and therefore the fiber $^{g'}H$ is not connected in $(G \square H) - S$. (Recall that every fiber meets C.) Therefore, $^{g'}H$ contains at least one edge from S, and we assign such an edge to (g', h). (Recall that, by definition, (g, h) has at most one neighbor in any H-fiber $^{g'}H$ where $g' \neq g$.) We proceed analogously for a neighbor (g, h') of (g, h) that lies in $^g H$.

Hence $|S| \geq \deg(g, h) \geq \delta(G \square H)$. $\qquad \square$

The proof of Theorem 5.5 is illustrated in Figure 5.3, where the bold edges represent edges from S.

5.3   Exercises

1. Let S be a separating set of a graph G. Show that then $S \times V(H)$ is a separating set of $G \,\square\, H$ for any graph H.

2. Find an infinite series of pairs of graphs G, H for which $\min\{\kappa(G)|H|, \kappa(H)|G|\} < \delta(G) + \delta(H)$.

3. Determine the connectivity of the grid graphs $P_m \,\square\, P_n$, $m, n \geq 2$.

4. Show that for any graph G, $\kappa(G \,\square\, K_2) = \min\{2\kappa(G), \delta(G) + 1\}$. Find an infinite family of graphs G such that $2\kappa(G) < \delta(G) + 1$ and an infinite family of graphs G for which $\delta(G) + 1 < 2\kappa(G)$.

5. Give a direct proof that for $n \geq 1$, $\kappa(Q_n) = n$.

6. Determine the edge-connectivity of $P_m \,\square\, P_n$, $m, n \geq 2$.

7. Prove that for any graph G, $\kappa'(G) \geq \kappa(G)$. Deduce from this and Exercise 5 that for any $n \geq 1$, $\kappa'(Q_n) = n$.

8. Show that $\kappa(G) = \kappa'(G)$ for any 3-regular graph G.

9. Let S be a disconnecting set of edges in a graph G. Show that for any graph H, the set of edges of $G \,\square\, H$ whose projections lie in S form a disconnecting set in $G \,\square\, H$.

10. [119] Let $G_1, \ldots, G_k$ be connected graphs on at least two vertices. Show that $\kappa'(\square_{i=1}^k G_i)$ is

$$\min\left\{ \sum_{i=1}^{k} \delta(G_i), \; \min_{1 \leq i \leq k} |G_1| \cdots |G_{i-1}| \kappa'(G_i) |G_{i+1}| \cdots |G_k| \right\}.$$

6 Subgraphs

The natural and easy concept of subgraphs has become most fruitful in graph theory. One reason for this is that numerous and important classes of graphs are characterized by forbidden subgraphs. Let us just recall that forbidden odd cycles as subgraphs characterize bipartite graphs, that the subgraphs K_5 and $K_{3,3}$ mark the origin of topological graph theory, and that we have used the characterization of outerplanar graphs by forbidden subgraphs in Chapter 4.

Another reason is that spanning subgraphs play a prominent role in the description of interesting problems. We have already seen this in Chapter 2, where the Hanoi graphs were described as spanning subgraphs of Cartesian powers of complete graphs. More important, and well known, is the role of spanning trees and spanning cycles (alias hamiltonian cycles; see Chapter 3) in combinatorial optimization.

It is hence natural to consider the subgraph structure of Cartesian products. Later we shall see that isometric and convex subgraphs play an important role for Cartesian products. In this chapter, however, we impose no restrictions of this kind. We first introduce nontrivial Cartesian subgraphs. Then, after some preliminary observations, we present two characterizations of nontrivial Cartesian subgraphs. One of them is stated in terms of a vertex-labeling condition and the other in terms of an edge-labeling condition. Finally, we briefly discuss basic trivial Cartesian subgraphs.

6.1 Nontrivial Subgraphs

Every graph G is a subgraph of $G \,\square\, H$ for any graph H. Hence every graph is a subgraph of many nontrivial Cartesian products. But it seems clear that such a containment (as a fiber) should be considered as a trivial one. This gives rise to the following definition.

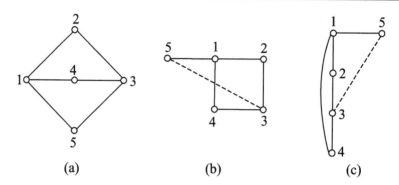

Figure 6.1. Trying to plot $K_{2,3}$: (a) complete bipartite graph $K_{2,3}$; (b) plot assuming 1234 is a square; and (c) plot assuming 1, 2, 3, and 4 are colinear.

A graph X is a *trivial Cartesian subgraph* if, as soon as X is a subgraph of a Cartesian product, X is a subgraph of one of the factors. More formally, X is a trivial Cartesian subgraph if $X \subseteq G \square H$ implies $X \subseteq G$ or $X \subseteq H$. If X is not a trivial Cartesian subgraph, it is called a *nontrivial Cartesian subgraph*. To shorten the presentation, we speak of a *trivial subgraph* and a *nontrivial subgraph*, respectively.

Clearly, any graph that is composite with respect to the Cartesian product is a nontrivial subgraph. However, from the definition of the Cartesian product we infer that K_3 is a trivial subgraph.

A *plotting* of a graph G is a drawing of G in the plane such that for any edge $e = uv$, u and v either have the same abscissa or they have the same ordinate; this is shown later in Figure 6.2(d). A plotting is *trivial* if all the vertices have either the same abscissa or the same ordinate. Now, Cartesian products are often drawn as plottings, that is, the vertices $V(G) \times V(H)$ are drawn in a rectangular grid. Then, as observed by Lamprey and Barnes [85], we easily infer that G is a nontrivial subgraph if and only if G has a nontrivial plotting. (See Exercise 1.)

For a simple example, consider the complete bipartite graph $K_{2,3}$; see Figure 6.1(a). In a possible plotting of it, the 4-cycle 1234 is either a square or a trivial subplotting. In the first case, vertex 5 must have the same abscissa or the same ordinate as 1, but then the edge between 3 and 5 violates the plotting condition; see Figure 6.1(b). Similarly, if the vertices 1, 2, 3, and 4 are on the same, say, vertical line, then either 5 is on the same line or has the same ordinate as 1. In the first case we have a trivial plotting, and in the second no plotting is possible. See Figure 6.1(c). Hence, $K_{2,3}$ is a trivial subgraph.

6.2 Characterizations

We now present the main theorem of this chapter. Here, a *properly colored path* means a labeled path for which any two consecutive vertices receive different labels (alias colors). The following result is due to Klavžar, Lipovec, Peterin, and Petkovšek [78, 80].

Theorem 6.1. [78, 80] *Let X be a connected graph on at least three vertices. Then the following statements are equivalent.*

(i)   *X is a nontrivial Cartesian subgraph.*

(ii) *$E(X)$ can be labeled with two labels such that on any induced cycle C of X that possesses both labels, the labels change at least three times while traversing the edges of C.*

(iii) *$V(X)$ can be labeled with k labels, $2 \le k \le |X| - 1$, such that the end vertices of any properly colored path receive different labels.*

Proof: (i) ⇒ (ii). Let X be a subgraph of $G \square H$ such that both $p_G X$ and $p_H X$ contain at least two vertices. Since X is connected, this implies that both $p_G X$ and $p_H X$ contain at least one edge. Define the labeling $\ell_E : E(X) \to \{1, 2\}$ as follows. For e an edge of X, set $\ell_E(e) = 1$ if $p_G(e)$ is an edge; otherwise, set $\ell_E(e) = 2$.

Let $C = x_1 x_2 \ldots x_k$ be an induced cycle of X that possesses labels 1 and 2. Suppose that labels change only twice on C. That is, the labels along C are $1, \ldots, 1, 2, \ldots, 2$, where $x_1 x_2$ is the first edge with label 1, and $x_r x_{r+1}$ is the first edge with label 2. Note that $r \ge 2$. Then the edges $x_1 x_2, \ldots, x_{r-1} x_r$ all lie in the same G-fiber, whereas the edges $x_r x_{r+1}, \ldots, x_{k-1} x_k$ belong to the same H-fiber. Hence, $x_1 = (g, h)$, $x_r = (g', h)$, where $g \neq g'$. Moreover, $x_k = (g', h')$, where $h \neq h'$. But then x_k is not adjacent to x_1 in $G \square H$, a contradiction.

(ii) ⇒ (iii). Let $\ell_E : E(X) \to \{1, 2\}$ be a labeling that uses both labels 1 and 2 such that on any induced cycle of X that possesses both labels, the labels change at least three times while traveling its edges. Let Y be the graph obtained from X by removing all edges e with $\ell_E(e) = 1$. Let $Y_1, \ldots, Y_k$ be the connected components of Y, and define $\ell_V : V(X) \to \{1, \ldots, k\}$ with $\ell_V(x) = t$, where $x \in Y_t$.

Since for at least one edge e of X we have $\ell_E(e) = 2$, at least two vertices of X receive the same ℓ_V label. Suppose Y is connected. Take a spanning tree of Y and add an arbitrary edge e with $\ell_E(e) = 1$ to the spanning tree. Then we find a cycle of X that violates the assumption that the labels change at least three times while traversing its edges. Hence, Y has at least two connected components, and therefore $2 \le k \le |X| - 1$.

Let $P = x_1 x_2 \ldots x_r$ be a properly colored (with respect to ℓ_V) path in X, and suppose that $\ell_V(x_1) = \ell_V(x_r)$. Since P is properly colored, any two consecutive vertices of P belong to different connected components of Y. Hence, for any edge $x_i x_{i+1}$ of P, we have $\ell_E(x_i x_{i+1}) = 1$. Now, since $\ell_V(x_1) = \ell_V(x_r)$, x_1 and x_r belong to the same connected component of Y. Thus, there exists an x_1, x_r-path R in Y. By the definition of ℓ_E, all edges of R are labeled 2. Combining P with R we get a cycle on which the ℓ_E-labels change only twice, a contradiction.

(iii) $\Rightarrow$ (i). Let now ℓ_V be a vertex labeling of X with k labels, $2 \le k \le |X| - 1$, such that the end vertices of any properly colored path receive different labels. We need to show that X is a nontrivial Cartesian subgraph.

Let Z be the spanning subgraph of X that is obtained by removing all edges $e = xy$ of X with $\ell_V(x) = \ell_V(y)$. Let $Z_1, \ldots, Z_r$ be the connected components of Z. Since $k \le |X| - 1$, there are vertices x and y of X with $\ell_V(x) = \ell_V(y)$. Then x and y must belong to different connected components of Z or we would have an x, y-path violating our assumption. Therefore $r \ge 2$.

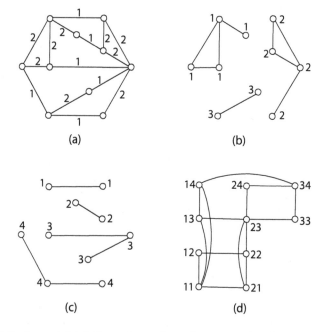

Figure 6.2. A nontrivial Cartesian subgraph. Parts (a)—(d) are described in the text.

To conclude the proof, we claim that X is a subgraph of $K_k \square K_r$. Let $V(K_n) = \{1, 2, \ldots, n\}$, and let $g : V(Z) \rightarrow V(K_r)$ be the natural projection. That is, $g(x) = i$, where $x \in Z_i$. Let $f : V(X) \rightarrow V(K_k \square K_r)$ be defined by $f(x) = (\ell_V(x), g(x))$. To show that f is injective and that it maps edges to edges is not too difficult, and is left as an exercise (Exercise 6).

Now, since the image of ℓ_V consists of at least two elements, X lies in at least two K_r-fibers, and the definition of g implies that X lies in at least two K_k-fibers, so X is a nontrivial subgraph. $\square$

Theorem 6.1 and its proof are illustrated in Figure 6.2. Figure 6.2(a) shows a graph G together with a 2-labeling of $E(G)$ that fulfills condition (ii) of the theorem. This is labeling ℓ_E from the proof. Figure 6.2(b) presents a vertex labeling obtained by removing edges labeled 1 in (a) and labeling the obtained connected components—these are the graph Y and the labeling ℓ_V. Then, in Figure 6.2(c), the graph Z is given as well as the function g. Finally, labelings from (b) and (c) are combined to obtain a plotting of G, as shown in Figure 6.2(d).

6.3 Basic Trivial Subgraphs

Let G be a graph on at least three vertices. Then G is a *basic trivial Cartesian subgraph* if it is a trivial subgraph that contains no proper trivial subgraph on at least three vertices.[1] As before, we speak here of *basic trivial subgraphs*.

The importance of basic trivial subgraphs lies in the fact that a trivial subgraph on at least three vertices is either a basic trivial subgraph or can be constructed from basic trivial subgraphs by two operations [86].

Lamprey and Barnes [86] listed a couple of examples of basic trivial subgraphs and an infinite family of such graphs were constructed by Klavžar, Lipovec, and Petkovšek [78]. (See also Exercise 7.) The first three graphs from this series are shown in Figure 6.3. It should be clear how the series continues. Note that the series starts with $K_{2,3}$.

Brešar [16] followed with a detailed study of basic trivial subgraphs and, among other results, constructed several infinite families of such graphs.

---

[1] Originally, Lamprey and Barnes [86] introduced these graphs as the graphs on at least three vertices that contain no proper *basic* trivial Cartesian subgraph. It was later shown that the definitions are equivalent [78].

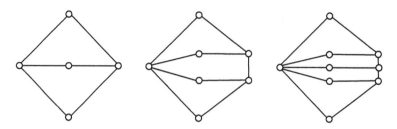

Figure 6.3. First three basic trivial Cartesian subgraphs.

6.4 Exercises

1. Show that a graph X is a nontrivial Cartesian subgraph if and only if X has a nontrivial plotting.

2. [85] Let G be a graph on at least three vertices with a cut vertex. Show that G is a nontrivial Cartesian subgraph by constructing an appropriate plotting.

3. [85] Let G be a graph on at least three vertices, and suppose that there exists a partition $\{A, B\}$ of $V(G)$ such that $A, B \neq \emptyset$ and the set of edges with one endpoint in A and the other in B form a matching. Show that G is a nontrivial subgraph by constructing an appropriate plotting.

4. Show that the graphs from Exercises 2 and 3 are nontrivial subgraphs by applying Theorem 6.1.

5. [78] Let G be a bipartite graph with radius 2, and suppose that G contains no subgraph isomorphic to $K_{2,3}$. Prove that G is a nontrivial Cartesian subgraph.

6. Complete the proof of Theorem 6.1 by showing that the function f is injective and that it maps edges to edges.

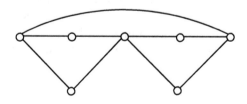

Figure 6.4. A basic trivial Cartesian subgraph.

7. Show that $K_{2,3}$ and the graph from Figure 6.4 are basic trivial Cartesian subgraphs.

8. The *direct product* $G \times H$ of graphs G and H has the same vertex set as their Cartesian product. The edge set of $G \times H$ consists of all pairs $[(g_1, h_1), (g_2, h_2)]$ of vertices with $[g_1, g_2] \in E(G)$ and $[h_1, h_2] \in E(H)$. Show that K_n, $n \geq 2$, is a trivial subgraph of the direct product; that is, $K_n \subseteq G \times H$ implies $K_n \subseteq G$ or $K_n \subseteq H$. (In fact, $K_n \subseteq G \times H$ implies $K_n \subseteq G$ *and* $K_n \subseteq H$.)

Part III

Graphical Invariants

7 Independence

Graphs have been used to model numerous problems in industrial, scientific, and social applications. Often independent sets of vertices play a crucial role for the solution of the ensuing problems. As a simple example, consider two multilane streets that meet at an intersection. The task is to design an efficient sequencing of a traffic signal that involves green and red directional arrows such that traffic can move through the intersection safely. We construct a graph that models this situation by letting each traffic lane be represented by a vertex. Two vertices are made adjacent in the graph if vehicles in the corresponding lanes would collide by simultaneously obeying a green arrow for their respective lanes. It is clear that any subset of the vertex set that induces a subgraph with no edges represents a set of lanes where traffic can safely respond to green arrows at the same time.

Here we study independence in Cartesian products. In particular, we present the best known bounds for the vertex independence number of a Cartesian product in terms of graphical invariants of the factors.

7.1 Definitions

A subset I of vertices in a graph G is said to be *independent* if the subgraph of G induced by I has no edges. It is an easy matter to find a maximal[1] independent set in a graph G. In fact, for any vertex u of G we can start with the set $\{u\}$ and repeatedly add a new vertex to the set as long as we maintain the property of independence. This

---

[1] We follow common practice in mathematics, and use the adjective "maximal"—in the context of a specific property—to describe a set that has the property such that no set containing it as a proper subset also has the property. The modifier "maximum" indicates a set of largest cardinality among all sets having the property. The use of the words minimal and minimum is similar.

greedy algorithm will produce an independent dominating set[2] (see Exercise 2) whose cardinality depends on the order in which the vertices of G are considered for inclusion. Its cardinality will lie somewhere between the *independent domination number* $i(G)$ and the *independence number* $\alpha(G)$, which are defined to be the minimum and maximum, respectively, of a maximal independent set of G. These values can be drastically different for a given graph. For example, if T is the star $K_{1,n+1}$, then $\alpha(T) - i(T) = n$. However, there are large classes of graphs for which the above algorithm always finds an independent set of maximum cardinality. (See Exercise 4.)

The central problem in this area is to compute, or at least to bound, the independence number. It is an NP-complete problem to decide if a graph G has an independent set of cardinality at least n if n is an input to the problem. For this reason, much of the research to date on the independence number of a Cartesian product concerns finding upper and lower bounds. In Sections 7.2 and 7.3 we consider lower and upper bounds, respectively, for $\alpha(G \,\square\, H)$. A graphical invariant that is useful for expressing some of these bounds is the *k-independence number* of G. This is defined to be the largest cardinality, $\alpha_k(G)$, of a subset A of $V(G)$ such that A is the union of k independent subsets, some of which are allowed to be empty.[3]

Note that an independent set in a graph G induces a complete subgraph, a *clique*, in the complement $\overline{G}$ of that graph. Thus, computing $\alpha(G)$ is equivalent to computing $\omega(\overline{G})$, the *clique number* of $\overline{G}$.[4] Any problem involving independent sets can be converted to a problem about complete subgraphs in the complement.

7.2 Lower Bounds

The central idea that is used to produce lower bounds for the independence number of Cartesian products is the fact that if I is an independent subset of G, and J is independent in H, then $I \times J$ is independent in $G \,\square\, H$. It follows immediately that

$$\alpha(G \,\square\, H) \geq \alpha(G)\alpha(H).$$

However, we can always do better unless one of the two factors has no edges. If vertices remain in both of $G - I$ and $H - J$, then independent

---

[2]For the definition of dominating set see p. 82.

[3]The k-independence number of a graph is also the order of a largest induced subgraph that has a k-coloring.

[4]Another invariant closely connected to the independence number is the *vertex cover number* $\beta(G)$; see Exercise 3.

sets I' and J' can be selected in $G - I$ and $H - J$, respectively. This gives rise to a larger independent set, namely $(I \times J) \cup (I' \times J')$. Of course, this process can be repeated as long as vertices remain in both graphs.

By choosing singleton sets starting with I' and J' and repeating the above procedure, Vizing [111] proved the first lower bound.

Proposition 7.1. [111] *For any graphs G and H,*

$$\alpha(G \square H) \geq \alpha(G)\alpha(H) + \min\{|G| - \alpha(G), |H| - \alpha(H)\}.$$

Let us now formalize the above discussion by following the method of Klavžar [76]. We wish to describe a procedure that will return a "large" independent set in $G \square H$. Let $G_1 = G$, $H_1 = H$, $k = 1$, and $I_0 = \varnothing$. Repeat the following sequence of three steps as long as the third step does not produce a graph with no vertices.

Step 1: Choose independent sets, A_k of G_k and B_k of H_k.

Step 2: Let $I_k = I_{k-1} \cup (A_k \times B_k)$ and increase k by 1.

Step 3: Let $G_k = G_{k-1} - A_{k-1}$ and $H_k = H_{k-1} - B_{k-1}$.

The set $I = I_m$, for the first value m such that G_{m+1} or H_{m+1} is a graph with no vertices, is independent in $G \square H$. If, at each stage in the procedure, the sets A_k and B_k are chosen to be maximal independent in G_k and H_k, respectively, then it is clear that I is maximal independent in $G \square H$. The set I is called a *maximal diagonal set* of $G \square H$. Jha and Slutzki [75] proposed this algorithmic approach in which they required the independent sets chosen in Step 1 to be maximum independent sets of the factors. It is perhaps surprising that for certain graphs it might be better not to require that A_k and B_k have largest possible order. The following result of Klavžar [76] shows a case where using maximum independent sets in the diagonal procedure is always best possible.

Theorem 7.2. [76] *Let G and H be bipartite graphs with bipartitions V_1, V_2 and W_1, W_2, respectively. If $\alpha(G) = |V_1|$ and $\alpha(H) = |W_1|$, then*

$$\alpha(G \square H) = |V_1||W_1| + |V_2||W_2|.$$

Consider the example in Figure 7.1. The graph G_1 is bipartite with bipartition sets $V_1 = \{1, 2, 3, 6, 7, 8, 12, 14, 16\}$ and $V_2 = \{4, 5, 9, 10, 11, 13, 15\}$ as indicated by shading in the figure. The independence number is $\alpha(G_1) = 11$, and so G_1 does not satisfy the hypothesis of Theorem 7.2. If maximum independent sets are chosen for this pair of

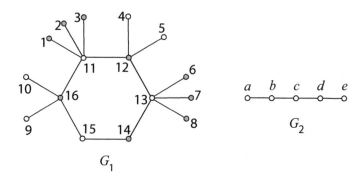

Figure 7.1. An example for computing $\alpha(G_1 \,\square\, G_2)$.

graphs in the above procedure, then the maximal independent set

$$I_2 = (\{1,2,3,4,5,6,7,8,9,10,14\} \times \{a,c,e\}) \cup (\{11,13,15\} \times \{b,d\}),$$

is found. However, $M = (V_1 \times \{a,c,e\}) \cup (V_2 \times \{b,d\})$ is also independent and $|M| = 41 > 39 = |I_2|$.

We now show that, in fact, $\alpha(G_1 \,\square\, G_2) = 41$. This demonstrates that it is not always best to use maximum independent sets in the procedure and also that the sufficient condition of Theorem 7.2 is not necessary. An application of Corollary 7.5 (ii) from Section 7.3 shows that $\alpha(G_1 \,\square\, G_2) \leq 43$. The reader is asked to verify this inequality in Exercise 7. Suppose first that $G_1 \,\square\, G_2$ has an independent set I such that $|I| = 43$. By Exercise 10, we may assume that G_1 has disjoint independent sets A and B such that $|A| = x$, $|B| = y$ and $43 = 3x + 2y$. Solving this Diophantine equation and using the fact that G_1 has order 16, we find that the only possible solution is $x = 11$ and $y = 5$. But G_1 has no independent set of order 11 whose complement is independent. Therefore, $\alpha(G_1 \,\square\, G_2) \neq 43$. In a similar way, we can show that no independent set of $G_1 \,\square\, G_2$ of cardinality 42 exists.

Not many lower bounds for the vertex independence number of an arbitrary Cartesian product are known. In the above example, where both graphs are of low order and one is an odd path, we were able to apply some very special techniques to calculate the number exactly. Hagauer and Klavžar [50] proved the following result, which holds when one of the factors is bipartite. The bound is expressed in terms of the 2-independence number of the other factor, which can be any graph.

For the next proposition, we recall that a *perfect matching* is a set of independent edges that meet every vertex of a graph.

Proposition 7.3. [50] *If G is any graph and H is bipartite, then*

$$\alpha(G \,\square\, H) \geq \frac{|H|}{2}\alpha_2(G),$$

and equality holds if H has a perfect matching.

Moreover, $\alpha(G \,\square\, H)$ can be computed exactly if G and H are paths or cycles. For the product of odd cycles, Hagauer and Klavžar [50] show that $\alpha(C_{2k+1} \,\square\, C_{2n+1}) = k(2n + 1)$ for $1 \leq k \leq n$. We leave the other cases as an exercise to the reader.

7.3 Upper Bounds

The most common method to bound the independence number of $G \,\square\, H$ from above is to partition the vertex set of one of the two factors, say H, into subsets that induce the subgraphs $H_1, H_2, \ldots, H_k$. If each of these subgraphs has a certain property, then an upper bound for $\alpha(G \,\square\, H)$ can be derived from the independence number of each $G \,\square\, H_i$. Klavžar [76] showed that all previously known upper bounds fit into this framework.

Theorem 7.4. [76] *Let $\{V_1, V_2, \ldots, V_k\}$ be a partition of the vertex set of a graph H, and let H_i be the subgraph of H induced by V_i. Then for any graph G,*

$$\alpha(G \,\square\, H) \leq \sum_{i=1}^{k} \alpha(G \,\square\, H_i).$$

For example, suppose that H has a perfect matching. That is, H has even order $2k$ and a set $\{e_1, e_2, \ldots, e_k\}$ of independent edges. The graph kK_2, the disjoint union of k edges, is a spanning subgraph of H. We observe that any largest independent set of $G \,\square\, K_2$ can be realized in the form $(I_1 \times \{1\}) \cup (I_2 \times \{2\})$, where I_1 and I_2 are disjoint independent sets of G. It follows (see Exercise 5) that

$$\alpha(G \,\square\, H) \leq \sum_{i=1}^{k} \alpha(G \,\square\, K_2) = k\alpha_2(G).$$

Of course, many graphs do not admit a perfect matching, but it is always possible to partition $V(H)$ into subsets $V_1, V_2, \ldots, V_r$, each of which induces a complete subgraph of H. Suppose this is the case and assume that $|V_i| = n_i$. Again, an immediate application of Theorem 7.4 and the result of Exercise 5 show that for any G,

$$\alpha(G \,\square\, H) \le \sum_{i=1}^{r} \alpha_{n_i}(G).$$

By finding an appropriate way to partition $V(H)$, the upper bounds of Corollary 7.5 are also easily established. Let $\alpha'(G)$ denote the maximum number of independent edges in G. This invariant is known as the *matching number of G*.

Corollary 7.5. *For any graphs G and H,*

    (i)   [111] $\alpha(G \,\square\, H) \le \min\{\alpha(G)\,|H|, \alpha(H)\,|G|\}$;

    (ii)   [50] $\alpha(G \,\square\, H) \le \alpha'(H)\alpha_2(G) + (|H| - 2\alpha'(H))\alpha(G)$.

Let us now consider the conditions that must hold in order for the value in the first inequality of Corollary 7.5 to be attained. Let us assume that $\alpha(G \,\square\, H) = \alpha(G)\,|H|$ and that I is a maximum independent set in $G \,\square\, H$. The set I must contain exactly $\alpha(G)$ vertices from each fiber G^h. In addition, for each edge $h_i h_j$ of H, the independent sets $p_G(I \cap G^{h_i})$ and $p_G(I \cap G^{h_j})$ are disjoint. The following simple example illustrates this idea.

Let G be the 5-cycle with vertices $g_1, g_2, \ldots, g_5$ and edges $g_i g_{i+1}$ modulo 5, and let H be the 10-cycle with vertices $h_1, h_2, \ldots, h_{10}$ and edges $h_i h_{i+1}$ modulo 10. Then the maximum independent sets of G are the five sets defined by $A_i = \{g_i, g_{i+2}\}$ for $1 \le i \le 5$ with vertex subscripts computed modulo 5. We can construct I via the following function. Let $\varphi : V(H) \to \{A_1, A_2, A_3, A_4, A_5\}$ be defined as $\varphi(h_i) = A_i$, with the subscript on the independent set computed modulo 5. We choose I to be that subset of $V(G \,\square\, H)$ such that $p_G(I \cap G^{h_i}) = \varphi(h_i)$.

This function φ is a homomorphism from H into what Brešar and Zmazek [17] called the *independence graph*,[5] $\text{Ind}(G)$, of G. The vertex set of $\text{Ind}(G)$ is the set of all maximum independent sets of G, and two such sets are adjacent precisely when they are disjoint.

The proof of Theorem 7.6 is suggested by the above example and is left as an exercise.

Theorem 7.6. [59] *For any two graphs G and H,*

$$\alpha(G \,\square\, H) = \min\{\alpha(G)\,|H|, \alpha(H)\,|G|\}$$

if and only if there is a homomorphism of G to $\text{Ind}(H)$ *or a homomorphism of H to* $\text{Ind}(G)$.

---

[5]Hell, Yu, and Zhou [59] called the collection $\{A_i\}$ of maximum independent sets of G an *independent set cover of G with respect to H*.

When both graphs are connected and bipartite, the necessary and sufficient condition of Theorem 7.6 can be restated without reference to homomorphisms because there exists a homomorphism from a bipartite graph to another graph X if and only if X has at least one edge.

Corollary 7.7. [17] *If G and H are connected, bipartite graphs, then*

$$\alpha(G \,\square\, H) = \min\{\alpha(G)\,|H|, \alpha(H)\,|G|\}$$

if and only if $|G| = 2\alpha(G)$ or $|H| = 2\alpha(H)$.

7.4 Exercises

1. Suppose G is a graph such that $V(G)$ can be partitioned into n independent sets. Show that $\alpha(K_n \,\square\, G) = |G|$.

2. Show that an independent set in a graph G is a dominating set of G if and only if it is a maximal independent set.

3. A *vertex cover* of a graph G is a set of vertices such that every edge of G is incident with at least one vertex in the set. Let $\beta(G)$ denote the minimum order of a vertex cover of G. Prove that $\alpha(G) + \beta(G) = |G|$.

4. [36] Let H_n be a graph of order $5n$ constructed from the disjoint union of n copies of C_5 by repeating the following operation an arbitrary number of times. Select two different 5-cycles and add an edge joining a vertex from one cycle with a vertex from the other in such a way that none of the original 5-cycles has two consecutive vertices whose degree is more than two. Show that $i(H_n) = 2n = \alpha(H_n)$.

5. Show that $\alpha(G \,\square\, K_n) = \alpha_n(G)$ for any graph G.

6. Prove Corollary 7.5.

7. Show that $\alpha(G_1 \,\square\, G_2) \le 43$, where G_1 and G_2 are the graphs in Figure 7.1.

8. Prove Theorem 7.6.

9. Let H be a graph. Prove that

$$\alpha(P_6 \,\square\, H) = \min\{3|H|, 6\alpha(H)\}$$

if and only if H is bipartite or H has a pair of disjoint maximum independent sets.

10. [50] Let P_{2n+1} be the path with vertex set $v_1, v_2, \ldots, v_{2n+1}$ in the natural order, and let G be any graph. Show that there is a maximum independent set I of $G \,\square\, P_{2n+1}$ with the property that for some disjoint independent sets A and B of G,

$$I = (A \times \{v_1\}) \cup (B \times \{v_2\}) \cup (A \times \{v_3\}) \cup \cdots$$
$$\cup (B \times \{v_{2n}\}) \cup (A \times \{v_{2n+1}\}).$$

11. Let each of G and H be a path or a cycle. Compute $\alpha(G \,\square\, H)$.

8 Graph Colorings

The subject of graph colorings was born in 1852, when Guthrie asked whether every planar map could be colored using at most four colors in such a way that two regions that share a border of positive length receive different colors. In 1976, the first correct proof of this conjecture was established by Appel and Haken [4]. During the hundred years leading up to this proof, a body of theory was developed in the language of graph theory concerning partitions of the vertex set of a graph.

In the first section of this chapter, we give some basic facts about the chromatic number. Among other things, we show that the chromatic number of a graph G can be expressed as the smallest integer m, such that $G \,\square\, K_m$ has an independent set of order m. We have seen in Chapter 7 that the independence number of Cartesian products is difficult to compute. In contrast, the computation of the chromatic number of a nontrivial Cartesian product can be reduced to its computation on the factors.

A coloring of a graph is equivalent to partitioning its vertex set into independent sets. However, other conditions can be imposed on the subsets in the partition. These conditions give rise to the so-called generalized colorings that we study in the second section of this chapter.

In the concluding section, we briefly introduce one of the most important recent generalizations of the chromatic number, the circular chromatic number. For a more in-depth study of this invariant, we refer the reader to the comprehensive and well-written survey article by Zhu [122].

8.1   Vertex Colorings

A *vertex n-coloring* of a graph $G = (V, E)$ is a function c from G to the set $\{1, 2, \ldots, n\}$ such that if $uv \in E$, then $c(u) \neq c(v)$. Assuming that $V(K_n) = \{1, 2, \ldots, n\}$ it follows that for each i, the set of vertices of color i, that is, $c^{-1}(i)$, is an independent set. We say the partition $c^{-1}(1), c^{-1}(2), \ldots, c^{-1}(n)$ of V is *induced* by c. In this way a partition of the vertex set of a graph G is also referred to as a *coloring* of G. The smallest such n for which G has a vertex n-coloring is called the *chromatic number* of G and is denoted $\chi(G)$.

A more modern approach uses the notion of mappings. Recall from Chapter 1 that a weak homomorphism from one graph G to another graph H is a function $p : G \rightarrow H$ such that $uv \in E(G)$ implies that $p(u)p(v) \in E(H)$ unless $p(u) = p(v)$. In other words, under p, the image of an edge is an edge or a single vertex. The projection mappings of a Cartesian product onto the two factor graphs are weak homomorphisms. A *homomorphism* from G to H is a mapping $f : G \rightarrow H$ such that adjacent vertices in G are mapped to adjacent vertices in H. Thus we see that $\chi(G)$ is the smallest positive integer n such that there is a homomorphism from G to K_n. Note that if $\chi(G) = n$ and $c : G \rightarrow K_n$ is a homomorphism, then every edge ij of K_n is the image of an edge in G. That is, there exists $x \in c^{-1}(i)$ and $y \in c^{-1}(j)$ such that $xy \in E(G)$. (See Exercise 1.) Such a coloring is called *complete*.

At the other extreme of the chromatic number is the *achromatic number*, $\psi(G)$, defined as the largest positive integer m such that there exists a complete coloring $a : G \rightarrow K_m$. For example, $\chi(C_8) = 2$ and $\psi(C_8) = 4$. Compare Exercise 7 and make use of the number of edges in C_8. The definition immediately implies that G must have at least $\binom{m}{2}$ edges, and yet this condition is far from being sufficient.

One concept that has been useful in studying colorings of graphs is that of color-critical graphs. The graph G is called *color-critical* or *χ-critical* if $\chi(H) < \chi(G)$ for every proper subgraph H of G. It is clear that every graph has a color-critical subgraph with the same chromatic number. Nontrivial Cartesian products are never color-critical. (See Exercise 3.)

As we saw in Chapter 7 for the independence number and will discover later in Chapter 11 with regard to the domination number, it is in general not possible to compute the value of an invariant on the Cartesian product from the values of the factors. However, as the next result of Sabidussi [103] demonstrates, this is not the case for the chromatic number.

Theorem 8.1. [103] *For any two graphs G and H,*

$$\chi(G \,\square\, H) = \max\{\chi(G), \chi(H)\}.$$

Proof: Since $G \,\square\, H$ has subgraphs isomorphic to G and H, the chromatic number of $G \,\square\, H$ is at least as large as $\max\{\chi(G), \chi(H)\}$. We may assume that $\chi(G) = k \geq \chi(H)$. Let $c_G : G \to \{1, 2, \ldots, k\}$ and $c_H : H \to \{1, 2, \ldots, \chi(H)\}$ be colorings. We construct a coloring c of $G \,\square\, H$ as follows. For an arbitrary vertex (g, h) of $G \,\square\, H$ we define

$$c(g, h) = c_G(g) + c_H(h) \pmod{k}.$$

For any edge $g_1 g_2$ of G and any vertex h of H, it follows immediately from the definition of c that

$$c_G(g_1) + c_H(h) \neq c_G(g_2) + c_H(h) \pmod{k}.$$

In other words, every fiber G^h is properly colored by c. Conversely, c_H is a coloring of H. Thus, for an edge $[(g, h_1), (g, h_2)]$ of the fiber ${}^g H$, we know that $c_H(h_1) \neq c_H(h_2)$. Since $c_H(h_1), c_H(h_2) \in \{1, 2, \ldots, k\}$,

$$c(g, h_1) = c_G(g) + c_H(h_1) \neq c_G(g) + c_H(h_2) = c(g, h_2) \pmod{k}.$$

Therefore, c is a coloring of ${}^g H$ as well, and it follows that

$$\chi(G \,\square\, H) = k = \max\{\chi(G), \chi(H)\}.$$

$\square$

See Figure 8.1 for an example illustrating this modular coloring of a Cartesian product.

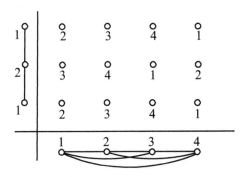

Figure 8.1. An example illustrating the proof of Theorem 8.1.

Not surprisingly, there is a relationship between $\alpha(G)$ and $\chi(G)$ for any graph G. Assume that G has order n and let $V_1, \ldots, V_{\chi(G)}$ be a coloring of G. Then, since each of these subsets is independent in G,

$$n = |G| = \sum_{i=1}^{\chi(G)} |V_i| \leq \sum_{i=1}^{\chi(G)} \alpha(G) = \alpha(G)\chi(G). \qquad (8.1)$$

We conclude that $\chi(G) \geq n/\alpha(G)$, or equivalently, $\alpha(G) \geq n/\chi(G)$. However, the complete k-partite graph $G = K_{n-(k-1),1,1,\ldots,1}$ of order n has $\chi(G) = k$ and $\alpha(G) = n - k + 1$, and so we see that (8.1) does not, in general, provide a good way to estimate $\chi(G)$ or $\alpha(G)$ if the other value is known.

Berge [10] provided the following connection between the chromatic number of a graph and the vertex independence number of a related Cartesian product.

Theorem 8.2. [10] *Let G be a graph of order n. The Cartesian product $G \square K_m$ has an independent set of order n if and only if G has an m-coloring.*

Proof: Suppose that $c : G \to K_m$ is a homomorphism, and let $V(K_m) = \{1, 2, \ldots, m\}$. The set $I = \{(x, c(x)) \mid x \in G\}$ has order n and is independent in $G \square K_m$ since c is a coloring.

Conversely, let A be an independent set in $G \square K_m$ such that $|A| = n$. Since any two vertices in the same K_m-fiber are adjacent, A can meet every K_m-fiber in at most one vertex. There are exactly n such fibers, and so A has to intersect every such fiber. It follows that for every $g \in G$, there is a uniquely determined i such that $(g, i) \in A$. We set $c(g) = i$. In other words, for $g \in G$, we have $c(g) = p_{K_m}(A \cap {}^g K_m)$. It is now easy to check that c is an m-coloring. $\square$

Corollary 8.3. *Let G be a graph of order n. The chromatic number of G is the smallest integer m such that $G \square K_m$ has an independent set of order n.*

For a given ordering $v_1, v_2, \ldots, v_n$ of the vertices of a graph G, the *greedy coloring* with respect to this ordering is the function g such that $g(v_1) = 1$, and for $i > 1$, $g(v_i)$ is the smallest positive integer not belonging to $g(N(v_i) \cap \{v_1, \ldots, v_{i-1}\})$. Regardless of the initial ordering, when a vertex u is encountered in the list, at most $\deg(u)$ colors are not eligible for $g(u)$. Hence, the largest possible value that can be assigned by g is $\Delta(G) + 1$, where $\Delta(G)$ denotes the maximum degree of G. Since g is a coloring, we have verified the bound in Theorem 8.4 for the chromatic number.

Theorem 8.4. *For any graph G, $\chi(G) \leq \Delta(G) + 1$.*

The bound of $\Delta(G) + 1$ of Theorem 8.4 is attained in any graph G that is either an odd cycle or a complete graph. The following result of Brooks [18] shows that the upper bound can be reduced by one in any other connected graph. The reader is asked to consider the proof of one case in Exercise 10.

Theorem 8.5. [18] *If G is any connected graph that is not an odd cycle or a complete graph, then $\chi(G) \leq \Delta(G)$.*

8.2   Generalized Colorings

As mentioned earlier in this chapter, a vertex coloring of a graph G can be considered as a partition of $V(G)$. Each subset in the partition must be independent; that is, it induces a subgraph with no edges. An obvious generalization is to require partitions such that each subset induces a subgraph of G having some other property of interest. We focus here on hereditary properties. (See [14] for a survey.) A *graph property* $\mathcal{P}$ is a class of graphs in which two isomorphic graphs are considered the same. Property $\mathcal{P}$ is said to be *hereditary* if whenever $G \in \mathcal{P}$ and H is a subgraph of G, then $H \in \mathcal{P}$ as well.

Let $\mathcal{P}$ be a hereditary property. A $\mathcal{P}$-*partition* of a graph G is a partition $V_1, \ldots, V_k$ of the vertex set of G such that $\langle V_i \rangle_G$, the subgraph of G induced by V_i, belongs to $\mathcal{P}$ for every i. The $\mathcal{P}$-*chromatic number* of G, denoted by $\chi_{\mathcal{P}}(G)$, is the smallest k for which G has such a $\mathcal{P}$-partition. This notion of *generalized colorings* was first studied by Hedetniemi in his Ph.D. dissertation [58]. For example, if $\mathcal{O}$ is the class of all finite graphs having no edges, then $\chi_{\mathcal{O}}(G) = \chi(G)$, the chromatic number of G defined earlier in this chapter. Following the notation of Borowiecki et al. [14], where $\mathcal{D}_1$ is the property of all forests, $\chi_{\mathcal{D}_1}(G)$ is the *vertex arboricity* of G. That is, $\chi_{\mathcal{D}_1}(G)$ is the minimum number of parts of a partition of $V(G)$ such that each part induces an acyclic subgraph.

For a hereditary property $\mathcal{P}$, the $\mathcal{P}$-chromatic number satisfies a natural and useful relationship on a graph and its subgraphs. Specifically, let H be a subgraph of G and let $V_1, V_2, \ldots, V_k$ be a $\mathcal{P}$-partition of G. Since $\mathcal{P}$ is hereditary, $V(H) \cap V_i$ induces a subgraph of H that belongs to $\mathcal{P}$ for each i. Thus, the $\mathcal{P}$-partition of G leads immediately to a $\mathcal{P}$-partition of H, and hence $\chi_{\mathcal{P}}(H) \leq \chi_{\mathcal{P}}(G)$.

Suppose that $\mathcal{P}_1$ and $\mathcal{P}_2$ are graph properties such that $\mathcal{P}_1 \subseteq \mathcal{P}_2$. Let G be a graph, and let $V_1, \ldots, V_k$ be a $\mathcal{P}_1$-partition of $V(G)$. It follows from the definitions that $V_1, \ldots, V_k$ is also a $\mathcal{P}_2$-partition of

$V(G)$. Hence, $\chi_{\mathcal{P}_2}(G) \le \chi_{\mathcal{P}_1}(G)$. For example, for every graph G, $\chi_{\mathcal{D}_1}(G) \le \chi_{\mathcal{O}}(G) = \chi(G)$. See Exercise 12 for a class of graphs where equality is attained.

Harary and Hsu [51] proved the following result, which is a generalization of Theorem 8.1.

Theorem 8.6. [51] *Let G and H be any graphs, and let $\mathcal{P}$ be a hereditary property. Then*

$$\chi_{\mathcal{P}}(G \,\square\, H) \ge \max\{\chi_{\mathcal{P}}(G), \chi_{\mathcal{P}}(H)\}.$$

Equality holds for all G and H if and only if the disjoint union and the Cartesian product of two graphs with property $\mathcal{P}$ have property $\mathcal{P}$.

Proof: By what was noted in the beginning of this section for a hereditary property $\mathcal{P}$, for any vertex h of H, $\chi_{\mathcal{P}}(G \,\square\, H) \ge \chi_{\mathcal{P}}(G^h) = \chi_{\mathcal{P}}(G)$, since G is isomorphic to the fiber G^h. Similarly, $\chi_{\mathcal{P}}(G \,\square\, H) \ge \chi_{\mathcal{P}}(H)$, and thus we have verified the inequality.

To prove the last statement we first assume that equality holds. If $\mathcal{P}$ is not closed under Cartesian products, then there exist two graphs H_1 and H_2 that are both in $\mathcal{P}$ but such that $H_1 \,\square\, H_2 \notin \mathcal{P}$. Consequently, $\chi_{\mathcal{P}}(H_1) = 1 = \chi_{\mathcal{P}}(H_2)$, but $\chi_{\mathcal{P}}(H_1 \,\square\, H_2) \ge 2$: a contradiction. If $\mathcal{P}$ is not closed under the disjoint union of graphs,[1] choose a pair of graphs $G_1, G_2 \in \mathcal{P}$ such that $G = G_1 + G_2$ is not in $\mathcal{P}$ and has the smallest possible order n. If $n = 2$, then $\chi_{\mathcal{P}}(2K_1) = 2$ and, since $\mathcal{P}$ is hereditary, it must also be the case that $\chi_{\mathcal{P}}(K_2) = 2$. Since we have assumed that equality holds in the statement of the theorem, $2 = \chi_{\mathcal{P}}(K_2 \,\square\, K_2)$. But in any $\mathcal{P}$-partition of $V(K_2 \,\square\, K_2)$, one of the parts must have at least two vertices. The subgraph induced by this part has property $\mathcal{P}$, contradicting what we know about $2K_1$ and K_2. Therefore, $n \ge 3$. We may assume that G_1 has order at least two and hence the order of G_2 is at most $n - 2$. Observe that the graph $K = G_2 + K_1$ is a proper subgraph of G, and G is a subgraph of $G_1 \,\square\, K$. According to the way we chose G, we see that $\chi_{\mathcal{P}}(K) = 1$ and, since $\mathcal{P}$ is hereditary, it follows that

$$\chi_{\mathcal{P}}(G_1 \,\square\, K) \ge \chi_{\mathcal{P}}(G) = 2 > 1 = \max\{\chi_{\mathcal{P}}(G_1), \chi_{\mathcal{P}}(K)\},$$

which contradicts our assumption that equality holds for all pairs of graphs with property $\mathcal{P}$. Hence, if equality holds in the statement of the theorem, then the disjoint union and the Cartesian product of two graphs with property $\mathcal{P}$ have property $\mathcal{P}$.

---

[1]For a precise definition of the disjoint union and properties with respect to the Cartesian product, see page 133.

Now, suppose that $\mathcal{P}$ is closed under union and Cartesian product and let G and H be any two graphs. If $m = \max\{\chi_{\mathcal{P}}(G), \chi_{\mathcal{P}}(H)\}$, then there are $\mathcal{P}$-partitions $U_1, \ldots, U_m$ of $V(G)$ and $V_1, \ldots, V_m$ of $V(H)$. For each k, $1 \le k \le m$, we define $W_k = \cup(U_i \times V_j)$, where the union is taken over all appropriate pairs of subscripts such that $k \equiv i + j$ (mod m). If this is not a $\mathcal{P}$-partition of $G \square H$, then some part W_k is not in $\mathcal{P}$. For $1 \le i \le m$, let $W_{k,i} = \{(g, h) \mid g \in U_i, (g, h) \in W_k\}$. Then $W_k = \cup_i W_{k,i}$. If $r \ne s$, $(g', h') \in W_{k,r}$, and $(g'', h'') \in W_{k,s}$, then (g', h') and (g'', h'') are not adjacent. This implies that $\langle W_k \rangle = \cup \langle W_{k,i} \rangle$. The property $\mathcal{P}$ is closed under disjoint union, and hence it follows that one of the subgraphs, say $\langle W_{k,t} \rangle$, does not belong to $\mathcal{P}$. If $W_{k,t} = \{(g_1, h_1), (g_2, h_2), \ldots, (g_n, h_n)\}$, then both $G' = \langle \{g_1, g_2, \ldots, g_n\} \rangle$ and $H' = \langle \{h_1, h_2, \ldots, h_n\} \rangle$ are in $\mathcal{P}$. But now we see that $\langle W_{k,t} \rangle$ is a subgraph of $G' \square H'$, and this contradicts the assumption that $\mathcal{P}$ is closed under Cartesian products since $\mathcal{P}$ is a hereditary property. $\square$

8.3   Circular Colorings

One of the most natural generalizations of the chromatic number of a graph, and a good approximation of it, is the circular chromatic number. This concept was introduced by Vince [110]. As perhaps implied by the name, it is used to model processes, such as the sequencing of traffic signals at a street intersection, which are to repeat cyclically. In such an example, each color class corresponds to a subset of the lanes of traffic that can safely proceed through the intersection simultaneously. The objective, of course, would be to minimize the number of distinct "green light" subsets in the cyclic sequence and thus to minimize the amount of stopping and starting by traffic.

For a positive number r, let C_r be a circle of circumference r. An r-*circular coloring* of the graph G is a function that assigns to each vertex of G an open arc of C_r having length 1 in such a way that adjacent vertices are assigned disjoint arcs. The *circular chromatic number*, $\chi_c(G)$, of G is the infimum of all values of r such that G has an r-circular coloring.

The statement that χ_c is a good approximation for χ is justified by the following result.

Theorem 8.7. [110] *For any graph G,*

$$\chi(G) - 1 < \chi_c(G) \le \chi(G).$$

There are several equivalent definitions of the circular chromatic number. We next describe how this concept was originally introduced by Vince [110].

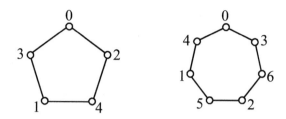

Figure 8.2. Colorings of C_5 (left) and C_7 (right).

For a positive integer k, let Z_k denote the cycle of order k with vertex set $\{0, 1, \ldots, k-1\}$ such that i is adjacent to $i-1$ and $i+1$ interpreted modulo k. If d is a positive integer such that $k \geq 2d$, then a (k, d)-*coloring* of a graph G is a function $c : G \to Z_k$ such that whenever u and v are adjacent vertices in G, then the distance between $c(u)$ and $c(v)$ in the graph Z_k is at least d. That is, $uv \in E(G)$ implies $d \leq |c(u) - c(v)| \leq k - d$. Then it turns out (Exercise 14) that

$$\chi_c(G) = \inf\{k/d \mid G \text{ has a } (k, d)\text{-coloring}\}. \tag{8.2}$$

Although it is not obvious, it can be shown that the infimum in Equation (8.2) can be replaced by the minimum. In other words, the infimum is actually assumed. Therefore, $\chi_c(G)$ is then a rational number. Figure 8.2 illustrates a $(5, 2)$-coloring of C_5 and a $(7, 3)$-coloring of C_7.

If c is a $\chi(G)$-coloring of G, then c is certainly a $(\chi(G), 1)$-coloring. This establishes the upper bound in Theorem 8.7.

Let Z_k^{d-1} denote the graph obtained from Z_k by adding an edge between any pair of vertices whose distance in Z_k is no more than $d-1$. A (k, d)-coloring of G is then simply a homomorphism from G to the complement of Z_k^{d-1}.

If G' is a subgraph of G, then an r-circular coloring of G can be restricted to $V(G')$, and the resulting function will be an r-circular coloring of G'. This immediately implies that $\chi_c(G') \leq \chi_c(G)$. Since the Cartesian product $G \square H$ has fibers isomorphic to G and others isomorphic to H, it follows that $\chi_c(G \square H) \geq \max\{\chi_c(G), \chi_c(H)\}$. We use the second of the two equivalent definitions above and present the proof of Zhu [121] to show the above inequality is actually always achieved.

Theorem 8.8. [121] *For graphs G and H, $\chi_c(G \square H) = \max\{\chi_c(G), \chi_c(H)\}$.*

Proof: Assume without loss of generality that $\chi_c(G) \geq \chi_c(H)$. From what was noted above, to prove Theorem 8.8, it suffices to show that $\chi_c(G \square H) \leq \chi_c(G)$. Suppose that $\chi_c(G) = k/d$, and let c_1 and c_2

be (k, d)-colorings of G and H, respectively. Let $f : G \square H \to Z_k$ be defined by $f(g, h) = c_1(g) + c_2(h)$ (mod k). The proof that f is a (k, d)-coloring is left as an exercise. $\qquad\qquad\qquad\qquad\square$

8.4 Exercises

1. Let G be a graph such that $\chi(G) = n$, and assume $c : G \to K_n$ is a homomorphism. Show that for $1 \le i < j \le n$ there exists $x \in c^{-1}(i)$ and $y \in c^{-1}(j)$ such that $xy \in E(G)$.

2. Let G and H be graphs with disjoint vertex sets. The *join* of G and H is the graph $G \vee H$ obtained from the union of G and H by adding edges making each vertex of G adjacent to each vertex of H. Show that
$$\chi(G \vee H) = \chi(G) + \chi(H).$$

3. Show that if G and H both have order at least two, then $G \square H$ is not color-critical.

4. Show that any greedy coloring g of G is a complete coloring. Conclude that the maximum value of g lies in the range from $\chi(G)$ to $\psi(G)$.

5. For any $n \ge 4$ find an ordering of the vertices of P_n such that the greedy coloring assumes a maximum value of 3 with respect to this ordering.

6. Show that for any graph G there is some ordering of $V(G)$ such that when the greedy coloring algorithm is applied to this ordering (as described just before Theorem 8.4), a coloring of value $\chi(G)$ is found.

7. Show that $\psi(P_8) = 4$. For each positive integer n, find a graph— in fact, a tree T—such that T has $\binom{n}{2}$ edges but $\psi(T) = 2$.

8. Let G be a graph with $\psi(G) = k \ge 3$. Show how to construct a complete $(k + 1)$-coloring of $G \square P_2$. This shows that $\psi(G \square P_2) \ge k + 1$. Show that this method gives the correct value of $\psi(P_4 \square P_2)$ but not the correct value of $\psi(P_8 \square P_2)$.

9. Let G be the bipartite graph $(\chi(G) = 2)$ of order $2n$ obtained by removing a perfect matching from $K_{n,n}$. Find an ordering of the vertices of G so that the greedy coloring assumes a maximum value of n.

10. Let G be a connected graph that is not regular. Prove that $\chi(G) \leq \Delta(G)$. (Hint: Choose an ordering of $V(G)$ whose last vertex has degree less than $\Delta(G)$ in such a way that the greedy coloring with respect to this ordering does not require more than $\Delta(G)$ colors.)

11. Show that $\chi_{\mathcal{D}_1}(K_n) = \lceil n/2 \rceil$.

12. Let $G = H_1 \square H_2$ where H_1 and H_2 are nontrivial, connected, bipartite graphs. Show that $\chi_{\mathcal{D}_1}(G) = \chi(G)$.

13. Let k be a positive integer. Consider the property $\mathcal{I}_k$ consisting of all graphs that do not contain a complete subgraph of order larger than $k + 1$. Show that

$$\chi_{\mathcal{I}_k}(G \square H) = \max\{\chi_{\mathcal{I}_k}(G), \chi_{\mathcal{I}_k}(H)\},$$

for all graphs G and H.

14. Verify Equation (8.2).

15. Complete the proof of Theorem 8.8 by showing that the proposed mapping of $V(G \square H)$ is a (k, d)-coloring.

Additional Types of Colorings

The focus in Chapter 8 was on partitioning the vertex set of a graph such that the parts of the partition induce subgraphs with predetermined properties. In this chapter, we are concerned with coloring models that pertain to optimization problems involving partitioning both the vertex and the edge set.

We begin with the introduction of the coloring number. It is an upper bound for the usual chromatic number and is related to colorings obtained by greedy algorithms. Then we study $L(2, 1)$-labelings that arise in channel assignment problems in communication networks. The chapter ends with a section on edge colorings.

9.1 List Colorings

To derive the bound in Theorem 8.4, it was sufficient to consider the greedy coloring with respect to an arbitrary ordering of the vertices. We now present an improvement to this bound by being more specific about the relationship between the ordering and the manner in which the greedy coloring assigns colors.

Let $V(G) = \{v_1, v_2, \ldots, v_n\}$, π be an arbitrary permutation of the set $\{1, 2, \ldots, n\}$, and $d_\pi(v_{\pi(i)})$ be the number of neighbors of the vertex $v_{\pi(i)}$ in $\{v_{\pi(1)}, \ldots, v_{\pi(i-1)}\}$. When the vertex $v_{\pi(i)}$ is colored by the greedy coloring with respect to the ordering $v_{\pi(1)}, \ldots, v_{\pi(n)}$, at most $d_\pi(v_{\pi(i)})$ colors are not available. This leads to the concept of the *coloring number* of G, col(G), defined as

$$\text{col}(G) = 1 + \min_\pi \max_i \{d_\pi(v_{\pi(i)})\},$$

where the minimum is computed over all permutations π of $\{1, 2, \ldots, n\}$. It follows immediately from the definition that

$$\chi(G) \leq \text{col}(G) \leq \Delta(G) + 1.$$

Proposition 9.1 establishes a natural connection between the coloring numbers of a Cartesian product and those of the factors. Its proof is left to the reader (see Exercise 2).

Proposition 9.1. *For any two graphs G and H,*

$$\mathrm{col}(G \,\square\, H) \le \mathrm{col}(G) + \mathrm{col}(H) - 1.$$

A notion closely related to the coloring number is list coloring. This was first introduced by Vizing [114] and independently by Erdös, Rubin, and Taylor [32].

Let k be a positive integer. Assigned to each vertex v of G is a subset (or list) $L(v)$ of cardinality k. The entire collection of subsets for all the vertices of G is called a list assignment of order k. We can, without loss of generality, assume that every $L(v)$ is a subset of $V(K_n)$ for some positive integer n. A *list coloring* of G corresponding to this list assignment is a homomorphism $c : G \to K_n$ such that $c(v) \in L(v)$ for every $v \in G$. The smallest integer $\chi_\ell(G)$ for which G has a list coloring corresponding to every possible list assignment of order $\chi_\ell(G)$ is the *list chromatic number* of G. Note in particular that the idea is to minimize k independently of the size of n.

It is easily seen that $\chi(G) \le \chi_\ell(G)$. First, if $k = \chi_\ell(G)$, let $L(v) = \{1, 2, \ldots, k\}$ for every $v \in G$. By the definition of the list chromatic number, there must be a homomorphism $c : G \to K_k$, and so G has a k-coloring. However, the chromatic number of G can also be strictly smaller than the list chromatic number. For example, consider the bipartite graph $K_{3,3}$ with bipartition $\{u_1, u_2, u_3\}$ and $\{v_1, v_2, v_3\}$ and the list assignment $L(u_1) = \{1, 2\} = L(v_1)$, $L(u_2) = \{1, 3\} = L(v_2)$, and $L(u_3) = \{2, 3\} = L(v_3)$ of order two. Neither choice of possible colors for u_1 from $L(u_1)$ can be extended to a coloring of the entire graph, and thus $\chi_\ell(K_{3,3}) > 2$.

We now prove a connection between the coloring number and the list chromatic number mentioned before.

Proposition 9.2. *For every graph G,* $\chi(G) \le \chi_\ell(G) \le \mathrm{col}(G) \le \Delta(G) + 1.$

Proof: The first and last inequalities have already been established. Suppose $V(G) = \{v_1, v_2, \ldots, v_n\}$ and let π be a permutation for which $\mathrm{col}(G) - 1 = \max_i\{d_\pi(v_{\pi(i)})\}$ in the definition of the coloring number of G. This implies that for every $1 \le i \le n$,

$$|N(v_{\pi(i)}) \cap \{v_{\pi(1)}, v_{\pi(2)}, \ldots, v_{\pi(i-1)}\}| \le \mathrm{col}(G) - 1.$$

If any list assignment of order $\mathrm{col}(G)$ is specified for G, then the above inequality guarantees that it is possible to find a coloring of G from these lists by considering the vertices in the order $v_{\pi(1)}, v_{\pi(2)}, \ldots,$ $v_{\pi(n)}$. □

The coloring number also plays a important role in an upper bound for the list chromatic number of the Cartesian product of two graphs. The result is due to Borowiecki, Jendrol, Král, and Miškuf [15]. We present the proof to illustrate the usefulness of fibers in Cartesian products.

Theorem 9.3. [15] *For any pair of graphs G and H,*

$$\chi_\ell(G \,\square\, H) \le \min\{\chi_\ell(G) + \mathrm{col}(H), \mathrm{col}(G) + \chi_\ell(H)\} - 1.$$

Proof: It is clearly enough to show that $\chi_\ell(G \,\square\, H) \le \chi_\ell(G) + \mathrm{col}(H) - 1$. To simplify the notation, let $k = \chi_\ell(G) + \mathrm{col}(H) - 1$ and assume that G has order p and H has order m. Let $h_1, h_2, \ldots, h_m$ be an ordering of the vertices of H such that for every $i > 1$,

$$|N(h_i) \cap \{h_1, h_2, \ldots, h_{i-1}\}| \le \mathrm{col}(H) - 1.$$

Suppose that L is a list assignment of order k for $G \,\square\, H$. Since the fiber G^{h_1} is isomorphic to G and since $k \ge \chi_\ell(G)$, there is a coloring c of G^{h_1} such that $c(g, h_1) \in L(g, h_1)$ for every vertex g of G. We now extend c inductively, and for simplicity we continue to call the new mapping c. Suppose $2 \le i \le p$ and that c has been extended to a list coloring of the subgraph induced by $G^{h_1} \cup G^{h_2} \cup \cdots \cup G^{h_{i-1}}$. For each $g \in G$ and for any $j < i$ such that $h_j \in N(h_i)$, we remove the color $c(g, h_j)$ if it belongs to the list $L(g, h_i)$. By the choice of the ordering of the vertices of H, this process will remove at most $\mathrm{col}(H) - 1$ colors from $L(g, h_i)$. Hence at least $k - \mathrm{col}(H) + 1 = \chi_\ell(G)$ remain in every list in the fiber G^{h_i}. Again, since this fiber is isomorphic to G, the coloring can be extended to G^{h_i}. In this way, a list coloring of $G \,\square\, H$ is produced. □

Borowiecki, Jendrol, Král, and Miškuf [15] also show that the bound of Theorem 9.3 is sharp.

9.2 $L(2, 1)$-labelings

Other types of graph colorings have received much attention by researchers due to their applications in designing communications networks. In these situations, a condition is imposed on colors allowed on pairs of vertices at distance two as well as on those that are adjacent. (See Calamoneri's [19] survey of the general problem.) Perhaps the best known of these is the one that goes by the name of $L(2, 1)$-labelings.

A function $L : V(G) \to \{0,1,2,\ldots\}$ is an $L(2,1)$-*labeling* of G if $|L(u) - L(v)| \geq 2$ whenever u and v are adjacent and $L(u) \neq L(v)$ if the distance between u and v is two. In many applications, it is desirable to label the graph so that the difference between the largest and smallest labels is minimized. Because labels can all be reduced by the same amount to produce a new $L(2,1)$-labeling, we can assume the smallest label is 0. Thus, most research has focused on the *span* of L, which is defined to be the maximum value in the range of L. The minimum span over all $L(2,1)$-labelings of G is denoted by $\lambda_{2,1}(G)$ and is called the $L(2,1)$-*labeling number* of G.

It is not surprising that to decide whether a graph has an $L(2,1)$-labeling whose span is at most k is an NP-complete problem [48]. Many of the published results thus attempt to relate $\lambda_{2,1}(G)$ to other graphical invariants of G. We list only a few of the more natural approaches in this chapter and summarize what is known about this graphical coloring as it relates to Cartesian products.

Since any $L(2,1)$-labeling of a graph is a coloring, $\chi(G) \leq \lambda_{2,1}(G)$. Griggs and Yeh [48] proved the following relationship between the chromatic and the $L(2,1)$-labeling number.

Theorem 9.4. [48] *For any graph G, $\lambda_{2,1}(G) \leq |G| + \chi(G) - 2$.*

Another invariant that has a direct connection to the $L(2,1)$-labeling number is the maximum degree Δ, since the neighbors of each vertex must receive distinct labels. Griggs and Yeh [48] found an infinite class of regular graphs G whose span is at least $\Delta^2 - \Delta$. Their conjecture that $\lambda_{2,1} \leq \Delta^2$ for all graphs of maximum degree Δ has not been settled and has motivated much research. This upper bound is the correct value for paths of order at least five and for all cycles. In general, however, Δ^2 is a poor bound for graphs having relatively large maximum degree. In Exercise 8, the reader is asked to show that $\lambda_{2,1}(K_n) = 2n - 2$, which is much smaller than $(n - 1)^2$.

Concerning Cartesian products, the $L(2,1)$-labeling number of $P_m \,\square\, P_n$, $P_m \,\square\, C_n$, and $C_m \,\square\, C_n$ are known for all values of m and n. See Schwarz and Troxell [105]. These values range from 5 to 8, depending on m and n. Figure 9.1 shows two small Cartesian product graphs with an $L(2,1)$-labeling of each. The coloring of $P_2 \,\square\, C_6$ is optimal, as the reader should verify. This also follows from a more general result given in Exercise 11. The situation with $C_4 \,\square\, C_4$ is left for Exercise 12.

By using an optimal $L(2,1)$-labeling of G, it is straightforward to verify the following result. On one of the G-fibers, use the labeling of G, and on the other fiber, shift the values by the appropriate amount.

Proposition 9.5. *For any graph G, $\lambda_{2,1}(G \,\square\, K_2) \leq 2\lambda_{2,1}(G) + 1$.*

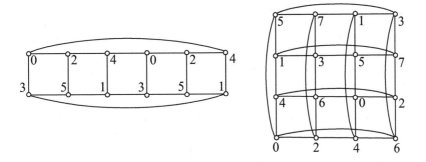

Figure 9.1. $L(2,1)$-labelings of $C_6 \,\square\, K_2$ (left) and $C_4 \,\square\, C_4$ (right).

Even though it does not claim to give exact values, the idea contained in the following lemma can prove useful in finding relative upper bounds for special Cartesian products.

Lemma 9.6. *For every graph G and every $n \geq 3$,*

$$\lambda_{2,1}(P_n \,\square\, G) \leq \lambda_{2,1}(K_3 \,\square\, G).$$

Proof: Let $V(K_3) = \{1,2,3\}$ and $V(P_n) = \{1,2,\ldots,n\}$. Suppose that $L : K_3 \,\square\, G \to \{0,1,\ldots,r\}$ is an optimal $L(2,1)$-labeling. We define a labeling L' of $P_n \,\square\, G$ in terms of L by $L'(k,g) = L(k \bmod 3, g)$. It is easy to check that this is in fact an $L(2,1)$-labeling of $P_n \,\square\, G$. $\square$

Georges, Mauro, and Stein [40, 41] investigated $L(2,1)$-labelings for Cartesian products of complete graphs. They show that for $2 \leq m \leq n$,

$$\lambda_{2,1}(K_m \,\square\, K_n) = mn - 1,$$

with the lone exception of $m = 2 = n$, in which case the value is 4. In addition, they computed exact values for higher Cartesian powers of complete graphs in special cases. One technique they used is related to a graphical invariant of the complement of the graph. The *path covering number* of a graph G is the smallest number, $c(G)$, of vertex disjoint paths that cover the vertex set of G. For example, G is traceable if and only if $c(G) = 1$. The connection between path covering numbers and $L(2,1)$-labelings is shown in the following result.

Proposition 9.7. [42] *Let G be a graph.*

  (i) $\lambda_{2,1}(G) \leq |G| + c(\overline{G}) - 2.$

  (ii) *If $\overline{G}$ is not traceable, then $\lambda_{2,1}(G) = |G| + c(\overline{G}) - 2.$*

9.3   Edge Colorings

Just as colorings of $G = (V, E)$ correspond to partitions of V into independent sets, so also we consider partitions of E into nonempty sets $E_1, E_2, \ldots, E_k$ such that for each i, E_i does not contain a pair of edges that share a vertex. Such a partition is called a *k-edge coloring* of G. The minimum such value of k is called the *chromatic index* of G and is denoted $\chi'(G)$. It is clear that each edge incident with a vertex of maximum degree must belong to a different part of the partition, and thus $\chi'(G) \geq \Delta(G)$. In 1964, Vizing [112] proved that any graph is of one of two types in terms of chromatic index.

Theorem 9.8. [112] *If G is any graph, then $\chi'(G) \leq \Delta(G) + 1$.*

The fundamental edge coloring problem is to decide whether a given graph G satisfies $\chi'(G) = \Delta(G)$ or $\chi'(G) = \Delta(G) + 1$. Graphs satisfying the first equation are said to be of *Class 1*; those satisfying the second are said to be of *Class 2*. It is an easy matter to show that even cycles and even complete graphs are of Class 1, whereas odd cycles and odd complete graphs are of Class 2. It is somewhat more difficult to prove that every bipartite graph is of Class 1.

With appropriately chosen Class 2 factors, the Cartesian product can be either Class 1 or Class 2. Mahmoodian's elegant proof of the following result shows that a Cartesian product is of Class 1 if at least one of the factors is of Class 1.

Theorem 9.9. [89] *If G is any graph with $\chi'(G) = \Delta(G)$, then $G \,\square\, H$ is of Class 1 for every graph H.*

Proof: Let $\Delta(G) = r$ and $\Delta(H) = s$. Then $\chi'(G) = r$ and $\Delta(G \,\square\, H) = r + s$. We show that $G \,\square\, H$ is a Class 1 graph by demonstrating how to color its edges with $r + s$ colors. Fix a coloring of $E(G)$ with colors $1, 2, \ldots, r$, and fix an edge coloring of H using colors $r+1, r+2, \ldots, r+\chi'(H)$. Color the edges of each fiber G^h exactly as the edges of G are colored. That is, assign the same color to the edge $[(g_1, h), (g_2, h)]$ as was assigned to the edge $g_1 g_2$. Similarly, for each vertex g of G, color edge $[(g, h_1), (g, h_2)]$ of fiber ${}^g H$ by using the same color as was used on edge $h_1 h_2$. If H is of Class 1, then this edge coloring uses $r + s$ colors. Otherwise, consider any edge, say $[(g_i, h), (g_j, h)]$, that is colored 1. The vertex (g_i, h) (respectively, (g_j, h)) is incident to at most $\Delta(H)$ edges in ${}^{g_i} H$ (respectively, in ${}^{g_j} H$), and since $\chi'(H) = \Delta(H) + 1$, there is at least one color from $r + 1, r + 2, \ldots, r + \chi'(H)$ missing in the edges at each of these end vertices of $[(g_i, h), (g_j, h)]$. By the way the edges of the fibers ${}^{g_i} H$ and ${}^{g_j} H$ were colored, the same color will be missing. Change the color of the edge $[(g_i, h), (g_j, h)]$ from 1 to this missing color. Repeating this process for each edge of $G \,\square\, H$ that was originally colored 1 yields an edge coloring of $G \,\square\, H$ using colors $2, 3, \ldots, r + s + 1$. Hence, $G \,\square\, H$ is of Class 1.   □

A graph G has a 1-*factorization* if $E(G)$ is the disjoint union of perfect matchings of G. As observed by Mahmoodian [89], Theorem 9.9 immediately implies the following result of Himelwright and Williamson [60]:

Corollary 9.10. [60] *If the graph G has a 1-factorization and H is regular, then $G \,\square\, H$ also has a 1-factorization.*

9.4 Exercises

1. Show that if T is a nontrivial tree, then $\mathrm{col}(T) = 2$. Show that $\mathrm{col}(G) = r + 1$ for an r-regular graph G.

2. Prove Proposition 9.1.

3. Show that $2 = \chi(K_{2,4}) < \chi_\ell(K_{2,4})$.

4. Prove that if H is a subgraph of G, then $\chi_\ell(H) \leq \chi_\ell(G)$.

5. Find the upper bound for $\chi_\ell(P_2 \,\square\, P_3)$ given by Theorem 9.3 and then determine the exact value.

6. Prove Theorem 9.4.

7. Compute the $L(2, 1)$-labeling number for the graphs P_6, C_6, K_8, $K_{3,3}$, and $P_3 \,\square\, P_3$.

8. Show that $\lambda_{2,1}(K_n) = 2n - 2$. Conclude that if G has order n, then $\lambda_{2,1}(G) \leq 2n - 2$.

9. Prove Proposition 9.5.

10. Show that $\lambda_{2,1}(K_n \,\square\, K_2) = 2n - 1$.

11. [48] Let G be a graph with maximum degree $\Delta \geq 2$. Show that $\lambda_{2,1}(G) \geq \Delta + 2$ provided that G contains a vertex of degree Δ that is adjacent to two other vertices of the same degree.

12. [73] Prove that the $L(2, 1)$-labeling of $C_4 \,\square\, C_4$ given in Figure 9.1 is optimal.

13. [48] Prove that $\Delta(T) + 1 \leq \lambda_{2,1}(T) \leq \Delta(T) + 2$ for every tree T.

14. [42] Show that if $\overline{G}$ is traceable, then $\lambda_{2,1}(G) \leq |G| - 1$.

15. Show that if G is a complete k-partite graph, then $\lambda_{2,1}(G) = |G| + k - 2$.

16. Let G be a regular graph of degree r. Prove that G has a 1-factorization if and only if $\chi'(G) = r$.

17. Use Exercise 16 to show that the Petersen graph is in Class 2.

10 Domination

The basic idea of domination theory can be illustrated most easily in the context in which it first appeared—the game of chess. A popular, recreational chess problem of the nineteenth century was to determine the minimum number of queens that could be placed on a (standard, 8×8) chessboard in such a way that each square that was not occupied by one of the queens was attacked by at least one queen. (A queen can move in any direction.) Variations of the original problem were to allow different size chess boards or to consider the same problem for a different chess piece.

To see that this chess problem is related to domination in graphs we construct a graph, say Q_8, that has 64 vertices, one for each square on the 8×8 board. Two vertices are made adjacent in Q_8 if a queen placed on one of the corresponding squares can attack the square represented by the other vertex. In this graphical model the "queens" problem becomes one of determining the number of vertices in a smallest subset of $V(Q_8)$ such that each vertex of the graph either belongs to the subset or is adjacent to at least one vertex in the subset. Posing the same problem for rooks instead of queens gives a much simpler problem, that of determining the minimum number of rooks whose placement will guarantee that each square is either occupied or attacked by one of the rooks. (A rook can move any number of unblocked squares vertically or horizontally.) The graph arising from this rooks problem is easily seen to be the Cartesian product, $K_8 \square K_8$.

10.1 Definitions and Notation

For two nonempty subsets D and X of $V(G)$, the set D *dominates* X if each vertex in $X - D$ is adjacent to a vertex in D. Equivalently, D dominates X if every vertex x of X belongs to the closed neighborhood $N[u]$ of at least one vertex $u \in D$; that is, $X \subseteq N[D]$. When $X = V(G)$, the set D is called a *dominating set* of G, and we say D dominates G. A dominating set of G is called a *minimal* dominating set if none of its proper subsets dominates G. The *domination number* $y(G)$ of G is the minimum cardinality of a dominating set of G. For example, for $2 \leq m \leq n$, $y(K_{m,n}) = 2$ and $y(P_n) = \lceil n/3 \rceil = y(C_n)$.

We see from the above definition that any dominating set D of a graph G must have a nonempty intersection with every closed neighborhood in G. That is, if $x \in G$, then $D \cap N[x] \neq \emptyset$. One way to obtain a lower bound for the domination number of G is to find a set A of vertices in G that has the property that if a_1 and a_2 are distinct members of A, then their closed neighborhoods are disjoint. It follows that the dominating set must have cardinality at least as large as that of A. Such a set A is called a 2-*packing* of G. The 2-*packing number* $\rho(G)$ is the maximum cardinality of a 2-packing. Thus we have the following proposition.

Proposition 10.1. *For any graph G, $y(G) \geq \rho(G)$.*

Although the domination number has been determined for many of the common classes of graphs (e.g., paths, cycles, and grid graphs with small widths), no efficient algorithm is known that accepts as input an arbitrary graph $G = (V, E)$ and produces as output a dominating set of minimum cardinality for G. In particular, the following related decision problem was shown to be NP-complete by Johnson [39]: Given a graph G and a positive integer k, does G have a dominating set D such that $|D| \leq k$?

Note that if u and v are nonadjacent vertices in G, then any dominating set of G is also a dominating set of the graph $G + e$, where $e = uv$. It follows immediately, then, that $y(G + e) \leq y(G)$. We can instead view this relationship by considering G as the graph obtained from $G + e$ by removing the edge e. A finite application of this process of deleting edges leads to the following result.

Proposition 10.2. *If G' is a spanning subgraph of G, then $y(G) \leq y(G')$.*

In particular, if T is a spanning tree of a connected graph G, then the domination number of G is bounded above by that of T. Although it is NP-hard to compute the domination number for arbitrary graphs, polynomial time algorithms have been developed for computing the domination number of certain classes of graphs (e.g., trees, block graphs, interval graphs, and permutation graphs). The first such algorithm was published in 1975 by Cockayne, Goodman, and Hedetniemi [23]. Their algorithm finds the domination number of a tree in time proportional to its order. By applying this algorithm to an arbitrary spanning tree of G, it is thus possible to obtain an upper bound

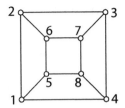

Figure 10.1. Graph H in example for algorithm.

for $y(G)$ that may be an improvement over the trivial upper bound of $|G|/2$ given by the result of Ore [96]. (See Exercise 7.) Of course, given the intractability of the dominating set problem, unless P=NP, it is not possible to *a priori* select a spanning tree whose domination number is equal to $y(G)$, even though such a spanning tree exists. (See Exercise 14.)

An upper bound for $y(G)$ can be found directly instead of applying an algorithm to a spanning subgraph. The entire vertex set $V(G)$ is a dominating set of G. Consider a specific ordering $S : v_1, v_2, \ldots, v_n$ of the vertices of G. Perform the following algorithm.

Step 1: Let $D = V(G)$ and $k = 1$.

Step 2: If $D' = D - \{v_k\}$ dominates G, then let $D = D'$.

Step 3: Increment k by 1. If $k \leq n$, repeat Step 2.

The set D that remains when the process terminates depends on the initial ordering S. Regardless of the ordering, it will be a minimal dominating set of G, but it is possible that $|D|$ is strictly larger than $y(G)$. In fact, for any minimal dominating set A of G, there is at least one ordering S of $V(G)$ such that the algorithm applied to S produces A. This ordering is found by simply listing the vertices of A, in any order, at the end of the sequence S. For example, consider the graph H in Figure 10.1.

If the algorithm is applied to the ordering $S_1 : 1, 3, 4, 5, 6, 7, 2, 8$, the resulting minimal dominating set produced is $D_1 = \{2, 8\}$, a minimum dominating set of H. However, if the ordering $S_2 : 1, 2, 3, 4, 5, 6, 7, 8$ is used, then the algorithm yields the minimal dominating set $D_2 = \{5, 6, 7, 8\}$. In principle then, the value of $y(G)$ could be computed by applying the above algorithm to every possible ordering of $V(G)$. Of course, this amounts to running the algorithm $n!$ times, hardly an improvement over simply checking all 2^n subsets to see which are dominating sets of G. However, a single application acting on a random ordering of $V(G)$ will yield an upper bound for $y(G)$.

10.2   Bounds for y in Terms of Other Invariants

It is a computationally difficult problem to compute the domination number of an arbitrary graph as indicated by the NP-completeness result above. However, many bounds for $y(G)$ are known, some in terms of other, more easily computed invariants for G and others in terms of graphical invariants that are of interest in their own right. In addition, when certain restrictions are placed on the structure of the graph G, it is possible to derive additional bounds for the domination number of G. Throughout this section, we assume that the graphs being considered are connected. See Exercise 3 for a justification of this assumption.

Assume that $c : V(\overline{G}) \to V(K_n)$ is a coloring of the complement of G; that is, c is a homomorphism of $\overline{G}$ to K_n. This coloring induces a partition $V_1, V_2, \ldots, V_n$ of $V(\overline{G}) = V(G)$ given by $V_i = c^{-1}(i)$ for each $1 \le i \le n$. In $\overline{G}$, each V_i is independent. Equivalently, if V_i is not empty, then the subgraph induced by V_i is complete in G. One vertex chosen from each of these n color classes of $\overline{G}$ constitutes a dominating set in G. This proves the following result.

Proposition 10.3. *For every graph G, $y(G) \le \chi(\overline{G})$.*

The bound given in Proposition 10.3 is sharp. This can be seen by letting G be $\overline{T}$, where T is any tree with at least one edge. It is easy to show that $y(G) = 2 = \chi(T) = \chi(\overline{G})$. However, there are graphs for which the difference $\chi(\overline{G}) - y(G)$ is arbitrarily large. (See Exercise 10.)

If x is any vertex of G, then it is clear that $V(G) - N(x)$ dominates G. Choosing x such that $\deg(x) = \Delta(G)$ proves that $y(G) \le |G| - \Delta(G)$. Also, since any vertex u of G dominates itself and $N(u)$, it follows that for any dominating set D of G we must have

$$|G| \le \sum_{u \in D} |N[u]| = \sum_{u \in D} (\deg(u) + 1) \le |D|(\Delta(G) + 1).$$

These two observations verify the following bounds for the domination number in terms of the order and the maximum degree of the graph.

Proposition 10.4. *For any graph G,*

$$\frac{|G|}{\Delta(G) + 1} \le y(G) \le |G| - \Delta(G).$$

Cycles and paths whose order is a multiple of three are examples of graphs for which the lower bound is attained. Note that in all such examples, the graph has a minimum dominating set that is also independent. In fact, more is true. If $\Delta(G) + 1$ divides $|G|$ and $y(G) = \frac{|G|}{\Delta(G)+1}$, then there must exist a minimum dominating set that is actually a 2-packing. Consider the special case when G is a hypercube. A 2-packing in G that also dominates G leads to the notion of a perfect binary code. This application is discussed in Chapter 11.

There are also upper bounds for $\gamma(G)$ as a function of the minimum degree $\delta(G)$. The best known is due to Arnautov [5]. The proof we give is that of Alon [3], and is one of the first uses of the probabilistic method in the area of domination theory.

Theorem 10.5. [5] *If G is a graph of order n with minimum degree $\delta \geq 1$, then*

$$\gamma(G) \leq n \frac{1 + \ln(\delta + 1)}{\delta + 1}.$$

Proof: Since $\delta \geq 1$, the number $p = \frac{\ln(\delta+1)}{\delta+1}$ lies in the interval $(0, 1)$. Construct a random subset X of $V(G)$ by selecting each vertex of G independently with probability p. For such an X, let $Y = V(G) - N[X]$. The set $D = X \cup Y$ is a dominating set of G. By its construction, it follows that the expected cardinality of X is np. Now consider an arbitrary vertex u in G. It belongs to Y if and only if the closed neighborhood of u fails to intersect X. This happens with probability $(1 - p)^{\deg(u)+1}$. Then we see that

$$(1 - p)^{\deg(u)+1} \leq (1 - p)^{\delta+1} \leq (e^{-p})^{\delta+1} = e^{-p(\delta+1)}.$$

The first inequality follows because $\deg(u) \geq \delta$ and $0 < 1 - p < 1$. For the second inequality, consider the continuous function f defined by $f(t) = e^{-t} + t - 1$. This function is increasing on the nonnegative real numbers and $f(0) = 0$; thus, $e^{-p} \geq 1 - p$.

Hence, the event $u \in Y$ has probability at most $e^{-p(\delta+1)}$. Computing the expected cardinality of D, we get

$$E(|D|) = E(|X|) + E(|Y|) \leq np + ne^{-p(\delta+1)} = n\frac{\ln(\delta + 1) + 1}{\delta + 1}.$$

Therefore, some dominating set has cardinality no larger than this expected value, and the theorem is established. $\square$

We close this chapter with two other results, similar to that of Ore [96] in Exercise 7, that provide an upper bound for the domination number of a connected graph if the minimum degree is suitably restricted. The proofs of these two theorems are much more difficult than Exercise 7 and will be omitted.

Part (i) of Theorem 10.6 is due to McCuaig and Shepherd [90]; part (ii) is due to Reed [99].

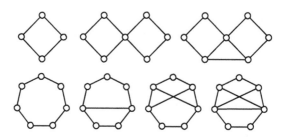

Figure 10.2. Seven exceptional graphs of Theorem 10.6.

Theorem 10.6. *Let G be a connected graph of order n. Then the following bounds hold:*

(i)   [90] *If $\delta(G) \geq 2$ and G is not one of the seven exceptional graphs from Figure 10.2, then $y(G) \leq 2n/5$.*

(ii) [99] *If $\delta(G) \geq 3$, then $y(G) \leq 3n/8$.*

Note that six of the exceptional graphs from Figure 10.2 are of order seven. The domination number of each of them is 3, which is larger than $14/5$.

10.3 Exercises

1. Show that five queens can be placed on a standard chessboard so as to occupy or attack each square. Show that the minimum number of rooks is eight. That is, show that $y(K_8 \square K_8) = 8$.

2. Show that in any graph G, a maximal independent set is also a dominating set, and thus $y(G) \leq \alpha(G)$.

3. Show that if a graph G has connected components $C_1, C_2, \ldots, C_r$, then $y(G) = \sum_i y(C_i)$.

4. Show that for $n \geq 3$, $y(P_n) = y(C_n) = \lceil n/3 \rceil$. Use this and Proposition 10.2 to prove that if G is a graph of order n that has a hamiltonian path or hamiltonian cycle, then $y(G) \leq \lceil n/3 \rceil$.

5. [70] Show that $y(P_2 \square P_n) = \lceil (n+1)/2 \rceil$.

6. Show that for a positive integer n, $y(P_2 \square C_n) = \lceil n/2 \rceil$ if $n \not\equiv 2$ (mod 4) and $y(P_2 \square C_n) = \lceil (n+1)/2 \rceil$ if $n \equiv 2$ (mod 4).

7. [96] Show that if a graph G has no isolated vertices and D is any minimal dominating set of G, then $V(G) - D$ dominates G. Conclude that if G has order n, then $y(G) \le n/2$.

8. [37] Show that if T is a tree of order $2k$, then $y(T) = k$ if and only if each vertex of T is either a leaf or a vertex adjacent to exactly one leaf.

9. Show that the graph H of Figure 10.1 has no minimal dominating sets of cardinality three.

10. Let G_2 be the graph of order ten formed from four copies of K_3 as follows. Identify one vertex from each of two of the copies of K_3 and call that vertex a. Do the same from the other two copies and call that vertex b. Add an edge joining a and b. See Figure 10.3. Show that $y(G_2) = 2$ but that $\chi(\overline{G_2}) = 4$. Generalize this construction to build a graph G_n of order $4n + 2$ such that $y(G_n) = 2$ and $\chi(\overline{G_n}) = 2n$.

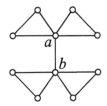

Figure 10.3. Graph G_2 for Exercise 10.

11. Let M be a maximal matching (that is, a set of independent edges maximal with respect to this property) in a connected graph G. Show that $y(G) \le 2|M|$.

12. Let M be a matching of maximum cardinality in a connected graph G. Show that $y(G) \le |M|$.

13. Show that for every edge e of a graph G, $y(G) \le y(G - e) \le y(G) + 1$. Show that for any vertex v of G, $y(G) - 1 \le y(G - v)$.

14. Let G be a connected graph. Show that G has a spanning tree T such that $y(T) = y(G)$. In fact, show that if D is any minimum dominating set of G, then there is a spanning tree T of G such that D is a dominating set of T.

15. Let G be a connected graph with diameter two. Show that $y(G) \le \delta(G)$.

11 Domination in Cartesian Products

In this chapter, we continue our study of domination, focusing on domination of Cartesian products. When trying to compute a fixed invariant of graphs, it is often useful to know that the graphs being considered are from a known class. The structure of the graphs in that class can be used to deduce certain bounds for the invariant. For example, the vertex independence number of a general graph G of order n can assume any value from 1 to n, but if only bipartite graphs are being considered, then the independence number will be at least $n/2$. In a similar way, the knowledge that a graph is the Cartesian product of two nontrivial graphs can be used to conclude certain things about the domination number.

Section 11.1 explains the important connection between domination and its application to coding theory. The famous conjecture by Vizing [113], which has motivated much of the recent research in the area of domination of Cartesian products, is presented in the second section. Some of the leading researchers in domination theory consider this conjecture to be the most important problem in the area. Several lower bounds for the domination number of a Cartesian product that were derived in attempts to settle this conjecture are proved in Section 11.3.

11.1 Binary Codes

Recall that the vertex set of the n-cube Q_n consists of all n-tuples of the form $(b_1, b_2, \ldots, b_n)$, where each $b_i \in \{0, 1\}$. For ease of reference, we follow the practice of denoting this vertex by b and shortening it to $b_1 b_2 \ldots b_n$. Note that if b and c are two vertices in Q_n, then the distance between them is given by the number of coordinates i for which $b_i \neq c_i$. See Chapter 2, Exercise 5, and the remark after Lemma 12.2. This distance is the Hamming distance between b and c. Thus, a subset D of $V(Q_n)$ is a dominating set of Q_n if and only if for every u of Q_n, the Hamming distance between u and some member of D is at most one.

Information is to be communicated from a sender to a receiver. The information, whether it is text, numerical data, a picture, or another form, is first encoded by translating it into some finite number of sequences of 0's and 1's such that each sequence has the same length. Each of these sequences is called a *codeword*. The sequence of codewords is then transmitted through some medium called a communication channel. When a codeword is received at the end of the channel, it must be decoded by reversing the encoding process to recover meaningful information. There is no attempt to encrypt the information so as to keep it secret. However, often there is interference or noise that affects the transmission and alters some of the bits of codewords. By allowing only certain codewords to be transmitted, it is sometimes possible to detect and even correct some of these transmission errors.

A *binary code of length n* is any subset C of vertices from Q_n. For example, let $C = \{1111, 1100, 0011\}$ with $n = 4$. When a word w of length 4 is received, it seems reasonable to decode it as the codeword in C whose Hamming distance from w is a minimum. Thus, if w is $0010, 0001$, or 0011, it would be assumed that one bit had been altered by noise in the first two cases and no transmission errors occur in the last case. The receiver would assume that 0011 had been transmitted and so w would be decoded as 0011. However, if $w = 0111$ or 1011, then the Hamming distance between w and each of the two codewords $1111, 0011 \in C$ is 1. In this case no unique decoding can be done based on the assumption that at most one transmission error can occur. A different problem would exist if $w = 1010$, for example. In this case, it is clear that more than one transmission error was made in sending the original codeword, and so some other corrective action—perhaps retransmitting—would need to be undertaken. What is needed is for the code in Q_n to be both a dominating set and a 2-packing; that is, a *perfect code*. Unfortunately, Q_4 does not have such a dominating set. Now, let $n = 3$ and $C = \{000, 111\}$. It is easy to verify that C is a perfect code of length 3.

Surprisingly, the exact value of $y(Q_n)$ is known only for a very limited collection of values of n. For $n \geq 4$ the complete list of what is known is given as follows [53]: $y(Q_4) = 4$; $y(Q_5) = 7$; $y(Q_6) = 12$; and for $n = 2^k - 1$ or $n = 2^k$, $y(Q_n) = 2^{n-k}$.

11.2   Multiplicative Results

For some graphical properties (for example, vertex independence), if A_i is a subset of vertices in the graph G_i that has the given property, then $A_1 \times A_2$ has the same property as a subset of the vertex set of $G_1 \,\square\, G_2$. In the case of vertex independence, it follows immediately that $\alpha(G_1 \,\square\, G_2) \geq \alpha(G_1)\alpha(G_2)$. However, if both graphs have order at least two, then the (set) Cartesian product of minimum dominating sets from both factor graphs is not a dominating set of the Cartesian product. (See Exercise 2.) The study of how graphical invariants behave on graph products can often be conveniently communicated by using the following two definitions. If φ is a graphical invariant and $\otimes$ is a graph product, then we say φ is *supermultiplicative* (respectively, *submultiplicative*) on $\otimes$ if for every pair of graphs G_1 and G_2,

$$\varphi(G_1 \otimes G_2) \geq \varphi(G_1)\varphi(G_2) \qquad (\text{resp., } \varphi(G_1 \otimes G_2) \leq \varphi(G_1)\varphi(G_2)).$$

Using this terminology, then, the vertex independence number α is supermultiplicative on $\square$. Vizing [111] studied dominating sets[1] in the context of Cartesian products and observed the following bound. It is established by considering a minimum dominating set of G in each fiber of the form G^h or a minimum dominating set of H in every fiber ${}^g H$.

Proposition 11.1. [111] *For all graphs G and H,*

$$\gamma(G \,\square\, H) \leq \min\{\gamma(H)\,|G|, \gamma(G)\,|H|\}.$$

In addition, Vizing attempted to prove a lower bound for the domination number of the Cartesian product of two graphs. He posed it as a question [111] and later as a conjecture [113].

Conjecture 11.2 (Vizing's Conjecture). [113] *The domination number γ is supermultiplicative on $\square$. That is, for any pair of graphs G and H,*

$$\gamma(G \,\square\, H) \geq \gamma(G)\gamma(H). \qquad (11.1)$$

To simplify the presentation, we say that *Vizing's Conjecture is true for the graph G* if Inequality (11.1) holds for every graph H. It is this intriguing conjecture that has provided much of the motivation for those studying domination on Cartesian products. Typically, it is not easy to compute the domination number of a Cartesian product; see

---

[1] The formal theory of domination in graphs seems to have had its origins in the 1950s, when a dominating set was called an *externally stable set*, and the domination number was the *coefficient of external stability*.

Exercise 5 for an exception. For most small examples for which the domination number can be computed, we find that not only is the inequality of Vizing's Conjecture true, but the proposed lower bound is much smaller than the actual value. This "evidence" in favor of the conjecture makes it compelling, if difficult.

First we present some of the bounds that have been established for the domination number $y(G \square H)$ for any pair of graphs G and H. Then we indicate several of the approaches that have been used to try to establish the conjecture.

11.3   Lower Bounds

Assume that G is a graph with vertices $g_1, g_2, \ldots, g_n$, and H has vertices $h_1, h_2, \ldots, h_m$. For a vertex (g, h) to dominate (g_i, h_j) in $G \square H$, either $g_i = g$ or $h_j = h$. Hence we see that if $A \subseteq V(G \square H)$ has fewer than $\min\{n, m\}$ vertices, there will necessarily be some vertex of the product that is not dominated by A. (Apply this to Exercise 1 in Chapter 10.) This proves the bound first published by El-Zahar and Pareek [30].

Proposition 11.3. [30] *For any G and H, $y(G \square H) \geq \min\{|G|, |H|\}$.*

Recall from Chapter 10 that $y(H) \geq \frac{|H|}{\Delta(H)+1}$. Jacobson and Kinch [70] established a lower bound that is reminiscent of this.

Proposition 11.4. [70] *For every pair of graphs G and H,*

$$y(G \square H) \geq \frac{|H|}{\Delta(H) + 1} y(G).$$

Proof: We assume G has order n and H has order m, as set out at the beginning of the section. Let D be a minimum dominating set of $G \square H$, and for each $1 \leq i \leq m$ let D_i consist of those vertices of the fiber, G^{h_i}, that belong to D. Consider the vertices of $G \square H$ to be arranged in an $n \times m$ grid so that the ith horizontal row is the fiber G^{h_i} and the jth vertical column is the fiber $g_j H$. We say a vertex (g, h_i) that is not in D_i is *horizontally undominated* by D if no vertex of D_i is adjacent to (g, h_i). Of course, such a vertex is adjacent to at least one vertex in $D \cap g H$. Denote by S_i those vertices in G^{h_i} that are horizontally undominated by D. It is clear that $|S_i| \geq y(G) - |D_i|$ because, if not, projecting $S_i \cup D_i$ onto G would produce a dominating set of G that has cardinality smaller than $y(G)$. But there must be enough edges

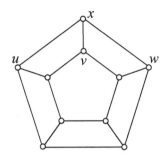

Figure 11.1. An example graph $G = C_5 \,\square\, K_2$.

incident with vertices in D to dominate $\cup_i S_i$. Therefore,

$$|D|\Delta(H) \geq |\cup_i S_i| \geq \sum_{i=1}^{m} (\gamma(G) - |D_i|) = m\gamma(G) - |D|.$$

The result follows by solving the inequality for $|D|$. $\square$

A lower bound for the domination number of the Cartesian product of two graphs that is easy to prove and that can be used to show that Vizing's Conjecture is true for several large classes of graphs depends on the 2-packing number. We illustrate the central idea in an example and leave the proof of the general result as an exercise. Let $G = C_5 \,\square\, K_2$, as shown in Figure 11.1, and let H be any graph. If we are interested in dominating only the vertices of the fiber ^xH by a subset $D \subseteq V(G \,\square\, H)$, then that part of D not belonging to $\{x, u, v, w\} \times V(H)$ can be disregarded. This is because $N[^xH] = N_G[x] \times V(H)$. If we now restrict the projection map p_H to $D \cap (N_G[x] \times V(H))$, we conclude that $|D \cap (N_G[x] \times V(H))| \geq \gamma(H)$.

Jacobson and Kinch [71] used this type of reasoning to establish the following lower bound.

Proposition 11.5. [71] *For any graphs G and H,*

$$\gamma(G \,\square\, H) \geq \max\{\gamma(G)\rho(H), \gamma(H)\rho(G)\}.$$

In general, this will not show that Vizing's Conjecture is true for a given graph. In particular, in our example above, $\rho(G) = 2 < 3 = \gamma(G)$. However, Meir and Moon [91] showed that $\rho(T) = \gamma(T)$ for every tree T. Thus Vizing's Conjecture is true for any tree. This fact was first proved by Barcalkin and German [6].

Corollary 11.6. [6] *If T is any tree, then $\gamma(T \,\square\, H) \geq \gamma(T)\gamma(H)$ for every graph H.*

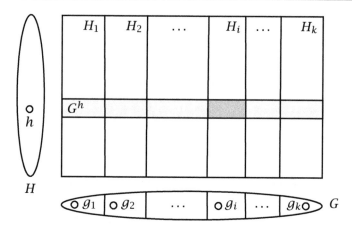

Figure 11.2. $G \square H$ for the proof of Theorem 11.7.

The only known lower bound for $y(G \square H)$ expressed in terms of the product of the domination numbers of both graphs was found by Clark and Suen [22]. Before their result, there was no known positive constant c such that $y(G \square H) \geq cy(G)y(H)$ for all pairs G and H. The proof we give is due to Goddard and Henning [44].

Theorem 11.7. [22] *For all graphs G and H,*

$$y(G \square H) \geq \frac{1}{2}y(G)y(H).$$

Proof: Let G and H be any graphs, and let D be a minimum dominating set of $G \square H$. Let $y(G) = k$, and let $\{g_1, \ldots, g_k\}$ be any minimum dominating set of G. Consider a partition $\{\pi_1, \ldots, \pi_k\}$ of $V(G)$ chosen so that $g_i \in \pi_i$ and $\pi_i \subseteq N[g_i]$ for each i. For each $1 \leq i \leq k$, let $H_i = \pi_i \times V(H)$. For a vertex h of H, the set of vertices $\pi_i \times \{h\}$ is called a cell. See Figure 11.2.

We say that such a cell is *vertically undominated* if none of its vertices belongs to $N[D \cap H_i]$; otherwise, we say it is *vertically dominated*. Let n_i be the number of cells in H_i that are vertically undominated. By considering the projection $p_H(D \cap H_i)$, it follows that $y(H) \leq |D \cap H_i| + n_i$. Now,

$$|D| + \sum_{i=1}^{k} n_i = \sum_{i=1}^{k} (|D \cap H_i| + n_i) \geq y(G)y(H). \qquad (11.2)$$

Since D dominates $G \square H$, if the cell $\pi_i \times \{h\}$ is vertically undominated, then $\pi_i \times \{h\}$ is dominated by $D \cap G^h$. Note that by the choice of the partition $\{\pi_1, \ldots, \pi_k\}$ (specifically that $\pi_j \subseteq N[g_j]$), the single vertex (g_j, h) dominates the cell $\pi_j \times \{h\}$ that is vertically dominated. If we let m_h denote the number of vertically undominated cells in G^h, then it follows by considering the projection of $D \cap G^h$ onto G that $\gamma(G) \leq |D \cap G^h| + (k - m_h)$. That is, $m_h \leq |D \cap G^h|$. Hence,

$$|D| = \sum_{h \in H} |D \cap G^h| \geq \sum_{h \in H} m_h = \sum_{i=1}^{k} n_i. \qquad (11.3)$$

By combining the inequalities in (11.2) and (11.3), we get

$$\gamma(G \square H) = |D| \geq \frac{1}{2}\gamma(G)\gamma(H).$$

$\square$

As indicated in Corollary 11.6, Vizing's Conjecture is true for any tree. Barcalkin and German [6] proved this as a corollary to their main result. They called a graph G *decomposable* if it has $\gamma(G)$ complete subgraphs whose vertex sets partition $V(G)$.

Theorem 11.8. [6] *If G' is a spanning subgraph of a decomposable graph G such that $\gamma(G') = \gamma(G)$, then Vizing's Conjecture is true for G'.*

This beautiful result, the first to verify the truth of the conjecture for a nontrivial class of graphs, was proved by first showing that Vizing's Conjecture is true for a decomposable graph and then applying the observation of Exercise 13. See Chapter 7 in Haynes, Hedetniemi, and Slater [57] for the proof of Theorem 11.8 as well as for a more detailed exposition of results concerning progress on the conjecture. The interested reader is invited to see Hartnell and Rall [54] for a result that generalizes Theorem 11.8. In addition, Hartnell and Rall [55, 56] improve some of the bounds on $\gamma(G \square H)$ presented in this chapter when additional restrictions are imposed on G or H.

11.4 Exercises

1. Let D be any dominating set of the Cartesian product $G \square H$, and let $A = p_G(D)$ and $B = p_H(D)$. Show that either $A = V(G)$ and B dominates H, or A dominates G and $B = V(H)$.

2. Assume that A dominates the graph G and that B dominates the graph H. Prove that $A \times B$ dominates $G \square H$ if and only if $A = V(G)$ or $B = V(H)$.

3. Show that $y(P_r \square P_s) \geq rs/5$.

4. Show that if Q_n has a perfect code C, then there exists a k such that $n = 2^k - 1$ and $|C| = 2^{n-k}$.

5. Show that if G is any graph of order n, then $y(G \square \overline{G}) = n$.

6. Use Proposition 11.4 to show that $y(C_{3r} \square H) \geq y(C_{3r})y(H)$ for any H.

7. Use Proposition 10.4 to show that if $n \geq m \geq 3$, then $y(C_n \square C_m) \geq y(C_n)y(C_m)$.

8. Prove Proposition 11.5. Conclude, for example, that Vizing's Conjecture is true for any cycle of order $3n$.

9. Suppose that T_1 and T_2 are trees such that for each vertex v, either $\deg(v) = 1$ or v is adjacent to exactly one vertex of degree one. Prove that each of T_1 and T_2 has domination number one-half its order and that $y(T_1 \square T_2) = y(T_1)y(T_2)$.

10. Let G and H be connected graphs of order n and m, respectively. For each vertex g of G, add a new vertex g' and the edge gg'. Call the new graph G'. Construct H' from H in a similar way. Show that $y(G') = n$, $y(H') = m$, and $y(G' \square H') = y(G')y(H')$.

11. Show that $y(G \square P_4) = 2y(G)$ implies $y(G \square C_4) = 2y(G)$.

12. Show that if G is a graph that has at least one edge, then $y(G \square P_3) > y(G) = y(G)y(P_3)$.

13. Assume that Vizing's Conjecture is true for some graph G and that G' is a spanning subgraph of G such that $y(G') = y(G)$. Show that the conjecture is also true for G'.

Part IV

Metric Aspects

12 Distance Lemma and Wiener Index

Metric graph theory abounds in applications. Let us just mention that it is applicable in such different areas as location theory, theoretical biology and chemistry, combinatorial optimization, and computational geometry.

In this chapter, we study basic connections between the Cartesian multiplication of graphs and metric graph theory. We consider the standard *shortest path distance* $d_G(u, v)$ between two vertices u and v of a connected graph G—that is, the number of edges on a shortest u, v-path. The main message of this chapter is that the Cartesian product is the natural product for this metric.

12.1 Distance Lemma

In this section, we show that the distance function is additive in the Cartesian product. As a consequence, it is usually not difficult to deduce the values of distance-based graph invariants from the corresponding values for the factors. We demonstrate it for the diameter, the radius, and the Wiener index.

For a path P of $G \square H$ consisting of a single edge e, we clearly have

$$|E(P)| = |E(p_G P)| + |E(p_H P)|,$$

because either $p_G P$ or $p_H P$ consists of a single vertex. If P is not a single edge, it may happen that two edges e and f of P have the same projection into one of the factors. Thus,

$$|E(P)| \geq |E(p_G P)| + |E(p_H P)|. \tag{12.1}$$

Lemma 12.1. *Let G and H be connected graphs, and let $(g, h), (g', h')$ be vertices of $G \square H$. Then*

$$d_{G \square H}((g, h), (g', h')) = d_G(g, g') + d_H(h, h').$$

99

Proof: Let P be a shortest $(g,h),(g',h')$-path in $G \square H$. By Inequality (12.1),

$$d_{G \square H}((g,h),(g',h')) \geq d_G(g,g') + d_H(h,h').$$

Furthermore, if R is a shortest g,g'-path in G, and S is a shortest h,h'-path in H, then $(R \times \{h\}) \cup (\{g'\} \times S)$ is a $(g,h),(g',h')$-path of length $d_G(g,g') + d_H(h,h')$. Hence,

$$d_{G \square H}((g,h),(g',h')) \leq d_G(g,g') + d_H(h,h').$$

By the associativity of the Cartesian product, Lemma 12.1 immediately extends to more than two factors. The extension is known as the *Distance Lemma*.

Lemma 12.2 (Distance Lemma). *Let G be the Cartesian product $\square_{i=1}^{k} G_i$ of connected graphs, and let $g = (g_1,\ldots,g_k)$ and $g' = (g'_1,\ldots,g'_k)$ be vertices of G. Then*

$$d_G(g,g') = \sum_{i=1}^{k} d_{G_i}(g_i,g'_i).$$

Let $G = \square_{i=1}^{k} K_{n_i}$ be a Hamming graph. Then the Distance Lemma implies that the distance between two vertices of G equals the number of coordinates in which they differ, a result we already know from Chapter 2, Exercise 6. We called this distance function the Hamming distance.

Despite its easy proof, the Distance Lemma is of utmost importance. Note first that it implies that

$$d_{G \square H}((g,h),(g',h)) = d_{G^h}((g,h),(g',h)).$$

In other words, $d_{G \square H}$ restricted to G^h is d_{G^h}.

This means that every shortest path in a G-fiber is also a shortest path in $G \square H$. Subgraphs with this property are called *isometric*. More precisely, a subgraph U of a graph G is *isometric* in G if $d_U(u,v) = d_G(u,v)$ for all $u,v \in U$. See Figure 12.1, where the filled vertices induce an isometric subgraph.

A particularly important classes of isometric subgraphs are formed by the isometric subgraphs of hypercubes and Hamming graphs. They are called, respectively, *partial cubes* and *partial Hamming graphs*.

We have thus observed that the G-fibers are isometric subgraphs of $G \square H$. By symmetry, the H-fibers are also isometric. In fact, an even stronger assertion holds. Every shortest $(G \square H)$-path between two vertices of one and the same fiber G^h or $^g H$ is already in that fiber. (See

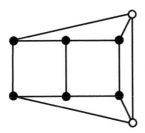

Figure 12.1. An isometric subgraph.

Exercise 1.) Such subgraphs are called convex. More precisely, a subgraph U of a graph G is *convex* in G if every shortest G-path between vertices of U is already in U. Note that the subgraph of Figure 12.1 is not convex.

Lemma 12.3. *Let G and H be connected graphs. Then all G-fibers and H-fibers are convex subgraphs of $G \square H$.*

12.2   Applications of the Distance Lemma

As a first application, we prove that intervals in Cartesian products induce boxes. (See Chapter 13 for a formal definition of a box.) This fact is a well-known part of the folklore, having been around at least since Mulder's book [94] was published in 1980.

For a connected graph G and $u, v \in G$, the *interval* $I_G(u, v)$ between u and v is defined as the set of vertices that lie on shortest u, v-paths; that is,

$$I_G(u, v) = \{w \in G \mid d(u, v) = d(u, w) + d(w, v)\}.$$

Proposition 12.4. *Let $X = G \square H$ be the Cartesian product of connected graphs G and H, and let (g, h) and (g', h') be vertices of X. Then*

$$I_X((g, h), (g', h')) = I_G(g, g') \times I_H(h, h').$$

Proof: Suppose $g'' \in I_G(g, g')$ and $h'' \in I_H(h, h')$. Then there is a shortest g, g'-path P in G containing g'':

$$P : g = g_1, g_2, \ldots, g_m = g'', g_{m+1} \ldots, g_n = g',$$

and a shortest h, h'-path Q in H containing h'':

$$Q : h = h_1, h_2, \ldots, h_r = h'', h_{r+1}, \ldots, h_s = h'.$$

Then

$$(g_1, h_1) \ldots (g_1, h_r)(g_2, h_r) \ldots (g_m, h_r) \ldots (g_n, h_r)(g_n, h_{r+1}) \ldots (g_n, h_s)$$

is a shortest $(g, h), (g', h')$-path containing $(g'', h'') = (g_m, h_r)$. Therefore, $I_G(g, g') \times I_H(h, h') \subseteq I_X((g, h), (g', h'))$.

To see that $I_X((g, h), (g', h')) \subseteq I_G(g, g') \times I_H(h, h')$, take an arbitrary shortest $(g, h), (g', h')$-path R. Then the Distance Lemma 12.2 implies that $p_G R$ is a shortest g, g'-path and $p_H R$ is a shortest h, h'-path. In particular, any vertex of R belongs to $I_G(g, g') \times I_H(h, h')$. $\square$

We proceed with applications of the Distance Lemma to the computation of graph invariants in Cartesian products. Recall that the *diameter* diam(G) of a connected graph G is defined as

$$\mathrm{diam}(G) = \max_{u, v \in G} d_G(u, v),$$

the *eccentricity* e(u) of a vertex u in G is defined as

$$e(u) = \max_{v \in G} d_G(u, v),$$

and the *radius* rad(G) is defined as

$$\mathrm{rad}(G) = \min_{u \in G} e(u).$$

Then, for connected graphs G and H, we easily infer that

$$\mathrm{diam}(G \,\square\, H) = \mathrm{diam}(G) + \mathrm{diam}(H)$$

and

$$\mathrm{rad}(G \,\square\, H) = \mathrm{rad}(G) + \mathrm{rad}(H).$$

In chemical graph theory, one of the most important graph invariants is the Wiener index.[1] Given a connected graph G, the *Wiener index* $W(G)$ is the sum of the distances between all pairs of vertices of G; that is,

$$W(G) = \frac{1}{2} \sum_{u \in G} \sum_{v \in G} d_G(u, v).$$

Computationally equivalent to the Wiener index is the *average distance* of a graph, which is the Wiener index divided by the number of pairs

---

[1] The Wiener index was introduced 1947 by Harold Wiener [117] for the investigation of the stability of chemical compounds. Nowadays, it belongs to molecular structure descriptors called topological indices that are used in theoretical chemistry for the design of chemical compounds with given physicochemical properties or given pharmacological and biological activities.

of vertices of a given graph G—that is, $W(G)/\binom{n}{2}$, where $n = |G|$. The average distance of a graph is an important graph invariant in several applications—for instance, in the design of communication networks.

By the Distance Lemma, it is not difficult to obtain a closed formula for the Wiener index of Cartesian products. It was first proved by Graovac and Pisanski [47] and obtained independently by Yeh and Gutman [120].

Proposition 12.5. [47, 120] *Let G and H be connected graphs. Then*

$$W(G \square H) = |G|^2 \cdot W(H) + |H|^2 \cdot W(G).$$

Proof: Set $V = V(G \square H)$, $V_G = V(G)$, and $V_H = V(H)$. Then

$$
\begin{aligned}
W(G \square H) &= \frac{1}{2} \sum_{(g,h) \in V} \sum_{(g',h') \in V} d_{G \square H}((g,h),(g',h')) \\
&= \frac{1}{2} \sum_{(g,h) \in V} \sum_{(g',h') \in V} (d_G(g,g') + d_H(h,h')) \\
&= \frac{1}{2} \sum_{g \in V_G} \sum_{h \in V_H} \sum_{g' \in V_G} \sum_{h' \in V_H} (d_G(g,g') + d_H(h,h')) \\
&= \sum_{g \in V_G} \sum_{g' \in V_G} \left(\frac{1}{2} \sum_{h \in V_H} \sum_{h' \in V_H} d_H(h,h') \right) + \\
&\quad\quad \sum_{h \in V_H} \sum_{h' \in V_H} \left(\frac{1}{2} \sum_{g \in V_G} \sum_{g' \in V_G} d_G(g,g') \right) \\
&= |G|^2 \cdot W(H) + |H|^2 \cdot W(G). \quad\quad\quad \square
\end{aligned}
$$

By the above result, the Wiener index of hypercubes can be obtained as follows. By definition, $W(Q_1) = 1$, and for $n \geq 2$,

$$W(Q_n) = W(Q_{n-1} \square K_2) = 2^{2n-2} \cdot 1 + 4 \cdot W(Q_{n-1}),$$

which can be solved to give

$$W(Q_n) = n2^{2(n-1)}.$$

12.3 Exercises

1. Show that every shortest $(G \square H)$-path between two vertices of one and the same fiber G^h or ^gH is already in that fiber.

2. Show that an arbitrary tree is a partial cube.

3. Show that the complete grid graphs $P_m \square P_n$ are partial cubes. More generally, prove that if G and H are partial cubes, then $G \square H$ is a partial cube as well.

4. A connected graph G is a *median graph* if

$$|I_G(u,v) \cap I_G(u,w) \cap I_G(v,w)| = 1$$

for any triple u, v, w of vertices in G. Show that the Cartesian product of two median graphs is a median graph as well.

5. For any $n \geq 2$, find a subgraph of Q_n that is isometric but not convex.

6. Show that among all trees with n edges, the Wiener index is minimized on the star $K_{1,n}$. In addition, determine $W(K_{1,n})$.

7. Show that among all trees with n edges, the Wiener index is maximized on the path with n edges.

8. Compute $W(P_m \square P_n)$, $m, n \geq 1$.

9. Show that a convex subgraph of a connected graph is connected.

13 Products and Boxes

In this chapter, we introduce boxes—those subgraphs of Cartesian product graphs that are products of subgraphs of the factors. We first prove the so-called Square Property and use it to characterize boxes. In the second section, we show that convex subgraphs of products are precisely boxes with convex projections, and we treat boxes related to distance centers and non-expansive mappings.

13.1 Boxes

Let U and V be subgraphs of G and H, respectively. Then $U \square V$ is a *box* of $G \square H$. Clearly, a subgraph W of a Cartesian product $G \square H$ is a box if and only if

$$W = p_G W \square p_H W.$$

The smallest nontrivial box is a square without diagonals. Such squares characterize the Cartesian product. This is amply supported throughout the book and illustrated by the next two results.

Lemma 13.1 (Square Property). *Let e and f be two adjacent edges of a Cartesian product that are in different fibers. Then there exists exactly one square in the product that contains both e and f. This square has no diagonals.*

Proof: Let e and f be adjacent edges of $G \square H$, where e is in a G-fiber and f is in an H-fiber. We can choose the notation such that $e = [(g,h),(g',h)]$ and $f = [(g,h),(g,h')]$; see Figure 13.1. By definition, the Cartesian product contains the vertex (g',h') and edges $e' = [(g,h'),(g',h')]$, $f' = [(g',h),(g',h')]$. Moreover, $efe'f'$ is a square and, again by the definition of the Cartesian product, neither the vertices (g,h) and (g',h') nor the vertices (g,h') and (g',h) are adjacent.

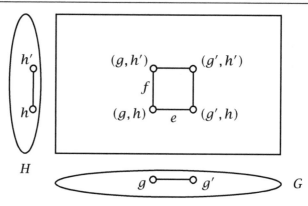

Figure 13.1. The Square Property.

Furthermore, any vertex not in G^h that is adjacent to (g',h) has coordinates (g',y), and any vertex not in ^gH that is adjacent to (g,h') has the coordinates (x,h'). Therefore, any vertex adjacent to both (g,h') and (g',h) that is different from (g,h) has the coordinates (g',h') and is thus unique.   □

Proposition 13.2. *A connected subgraph W of $G \square H$ is a box of $G \square H$ if and only if for any two adjacent edges e and f of W that are in different fibers, the unique square of $G \square H$ that contains e and f is also in W.*

Proof: Every connected box W clearly satisfies the conditions of the proposition, even if it is a connected subgraph of a fiber.

Conversely, let the condition of the proposition be satisfied. We have to show that $p_G W \square p_H W = W$. We first show that $p_G P \square p_H P \subseteq W$ for any shortest path P in W.

This is clearly so if P has only one edge. If it has more than one edge, let P be a uy-path with the first edge $e = uv$ and the last edge $f = xy$.

Suppose P has length 2; that is, e and f have a common vertex. If e and f are in different fibers, then $p_G e \square p_H f \subset W$ by assumption. If they are in the same fiber, say in G^h, then this is true because $G^h = p_G G^h \square p_H G^h = G \times \{h\}$.

We now proceed by induction with respect to the length of P, assuming that $p_G Q \square p_H Q \subseteq W$ for every proper subpath Q of P. Hence,

$$p_G (P - u) \square p_H (P - u) \subseteq W \quad \text{and} \quad p_G (P - y) \square p_H (P - y) \subseteq W.$$

Suppose e and f are in fibers with respect to one and the same factor, say with respect to G. Then

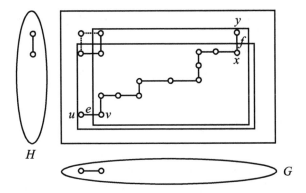

Figure 13.2. Situation from the proof of Proposition 13.2.

$$W \quad \supseteq \quad (p_G(P-u) \,\square\, p_H(P-u)) \cup (p_G(P-y) \,\square\, p_H(P-y))$$
$$= \quad (p_G(P-u) \,\square\, p_H P) \cup (p_G(P-y) \,\square\, p_H P)$$
$$= \quad (p_G(P-u) \cup p_G(P-y)) \,\square\, p_H P$$
$$= \quad p_G P \,\square\, p_H P.$$

On the other hand, if e and f are in fibers with respect to different factors, say $e \in G^u$ and $f \in {}^y H$, then

$$p_G P \,\square\, p_H P - ((p_G(P-u) \,\square\, p_H(P-u)) \cup (p_G(P-y) \,\square\, p_H(P-y)))$$

consists of the single vertex $(p_G(u), p_H(y))$ and the edges $p_G(u) \times p_H(f)$, $p_G(e) \times p_H(y)$. However, these three elements are contained in the unique square of $G \square H$ that contains the edges $p_G e \times p_H x$ and $p_G v \times p_H f$. Such a square exists by the Square Property since these two edges are adjacent and in different fibers; see Figure 13.2. This square is in W because

$$p_G e \times p_H x \in p_G(P-y) \,\square\, p_H(P-y) \subset W$$

and

$$p_G v \times p_H f \in p_G(P-u) \,\square\, p_H(P-u) \subset W.$$

Hence, $p_G P \,\square\, p_H P \subseteq W$.

To complete the proof, let $u' \in p_G W$ and $y'' \in p_H W$. Then there exist vertices $u, y \in W$ with $u' = p_G u$ and $y'' = p_H y$. In other words, $u = (u', p_H u)$ and $y = (p_G y, y'')$. Choosing a shortest path P from u to y in W and using the fact that $p_G P \,\square\, p_H P \subseteq W$, it is easily inferred that $(u', y'') \in W$.

Given any edge $u'v' \in E(p_G W)$ and any vertex $y'' \in p_H W$, a similar argument shows that $u'v' \square \{y''\} \in W$. By symmetry, this implies that $\{u'\} \square x'y' \in W$ if $u' \in p_G W$ and $x'y' \in E(p_H W)$. $\square$

Note that we did not assume that W was induced.

13.2 Metric Boxes

In the previous section, boxes of Cartesian products were introduced and characterized. Many interesting subgraphs of products are boxes; for example, convex subgraphs and distance centers are boxes.

If W is a convex subgraph of $G \square H$, then it is induced and satisfies the conditions of Proposition 13.2. Therefore it is a box, and, by Lemma 12.2, it is easy to see that it is convex if and only if both $p_G W$ and $p_H W$ are convex. We state this as the next result and leave the details of the proof for Exercise 1.

Proposition 13.3. *A subgraph W of a Cartesian product $G \square H$ is convex, if and only if it is a box*

$$p_G W \square p_H W,$$

where both projections $p_G W$ and $p_H W$ are convex.

So the convex subgraphs of Cartesian products are determined by the convex subgraphs of its factors. In contrast, the structure of isometric subgraphs of Cartesian products is much richer. Even in the simplest case, in hypercubes, the variety of isometric subgraphs is astonishing. For instance, trees, even cycles, median graphs, benzenoid graphs, and Cartesian products of partial cubes all belong to the class of partial cubes. See Section 18.3 for more on partial cubes.

In the rest of this section, we show how contractions of graphs give rise to isometric subgraphs and that contractions of isometric subgraphs of Cartesian products have fixed boxes.

A well-known special case is the center of a tree. It is defined as the set of its *central vertices*, where a vertex v is called *central* if $e(v) = \mathrm{rad}(T)$. It is easily seen that the center of a tree consists of one vertex or two adjacent ones, and that it is invariant under automorphisms. Thus, trees on at most two vertices coincide with their centers.

One of the standard proofs for the existence of centers in trees actually uses contractions. It can be briefly described as follows. If a tree T has at least three vertices, its center can be recursively constructed—see Exercise 2—by contracting all vertices of degree one with their neighbors. The resulting graph is a tree. If it has more than two vertices, the operation is repeated until the center is reached.

Note that the operation of identifying all vertices of degree one with their neighbors and fixing the others is a weak homomorphism that does not increase distances.

We next introduce a similar distance concept. The *distance center* $d(G)$ of a connected graph G is defined as

$$d(G) = \{u \mid \sum_{v \in G} d_G(u, v) \text{ is minimum}\}.$$

Often we will not distinguish between $d(G)$ and $\langle d(G) \rangle$. Thus,

$$d(G \square H) = d(G) \square d(H);$$

that is, the distance center of a product is a box that projects onto the distance centers of the factors. (See Exercise 4.)

A mapping $f : G \to G$ with $d(u, v) \geq d(f(u), f(v))$ is called a *contraction*. If it is idempotent, that is, if $f(f(v)) = f(v)$ for all $v \in G$, then it is a *retraction* and $f(G)$ is called a *retract*.

We next wish to show that the distance center of a retract of a Cartesian product graph G is a box in G. To this end, we first prove the following lemma, which holds for all graphs (not only for products).

Lemma 13.4. *Let H be a retract of a graph G and let*

$$C = \{g \in G \mid \sum_{h \in H} d_G(g, h) \text{ is minimum}\}.$$

Then H is isometric and $d(H) = C$.

Proof: Let $f : G \to H$ be a retraction. Since f is a contraction, H is isometric. Indeed, let u and v be vertices of H. Clearly, $d_G(u, v) \leq d_H(u, v)$. Now let P be a shortest u, v-path in G. Then $f(P)$ is a walk in H between $f(u) = u$ and $f(v) = v$. Therefore, $d_H(u, v) \leq d_G(u, v)$.

To show that $d(H) = C$, we first show that $d(H) \subseteq C$. For any $x \in C$, $f(x) \in H$, and since H is isometric and each vertex of H is fixed by f,

$$\sum_{h \in H} d_G(f(x), h) = \sum_{h \in H} d_G(f(x), f(h)) = \sum_{h \in H} d_H(f(x), f(h))$$
$$\leq \sum_{h \in H} d_G(x, h).$$

The inequality here must actually be an equality because $x \in C$, and therefore, by the definition of C, the sum $\sum_{h \in H} d_G(x, h)$ is as small as possible. It follows that $f(x) \in C$, and since $f(x) \in H$, we also infer that $f(x) \in d(H)$.

Suppose $w \in d(H)$. Then

$$\sum_{h \in H} d_G(f(w), h) = \sum_{h \in H} d_H(f(w), h) = \sum_{h \in H} d_H(w, h),$$

and hence $w = f(w) \in C$. Thus, we have shown that $d(H) \subseteq C$.

We next show that $C \subseteq d(H)$. Consider any $u \in C$. By the above, $f(u) \in d(H)$ and

$$\sum_{h \in H} d_G(f(u), h) \le \sum_{h \in H} d_G(u, h).$$

If $u \notin d(H)$, then $d_G(u, f(u)) > 0$ because $f(u) \in H$. But this implies $0 = d_G(f(u), f(u)) < d_G(u, f(u))$ and also

$$\sum_{h \in H} d_G(f(u), h) < \sum_{h \in H} d_G(u, h),$$

which is impossible. Hence $u \in d(H)$, and so $C \subseteq d(H)$. $\square$

We are now ready for the next result. It is due to Feder [33].

Theorem 13.5. [33] *Let H be a retract of a Cartesian product G. Then $d(H)$ is a box in G.*

Proof: Let $G = G_1 \square \cdots \square G_k$ and C be defined as in Lemma 13.4. Then $C = d(H)$ by this lemma. Furthermore, for any vertex $g = (g_1, \ldots, g_k) \in G$,

$$\sum_{h \in H} d_G(g, h) = \sum_{i=1}^{k} \sum_{h \in H} d_i(g_i, h_i),$$

where d_i denotes the distance in G_i. Let C_i be the set of g_i that minimize $\sum_{h \in H} d_i(g_i, h_i)$. Then $g \in C$ if and only if $g_i \in C_i$ for all i, which means that C is a box (in G). $\square$

Several important classes of graphs (e.g., median graphs—see Exercise 4 in Chapter 12) are retracts of Cartesian products. Using this fact and Theorem 13.5, we infer that their distance centers are boxes. In the case of median graphs, the distance centers are hypercubes.

We conclude this section with a generalization of Theorem 13.5 due to Tardif [108]. It has some similarity with the fixed point theorem for contractions of metric spaces.

Theorem 13.6. [108] *Let f be a contraction of a Cartesian product G. Then there exists a box B of G with $f(B) = B$, where B is the distance center of $f^k(G)$ for an appropriate power of f.*

The decisive part of the proof consists of showing the existence of an integer k with $f^{k+1}(G) = f^k(G)$. (See Exercises 5-9.)

13.3   Exercises

1. Show that a box B of a Cartesian product $G \square H$ is convex if and only if both $p_G B$ and $p_H B$ are convex.

2. Show that recursive contraction of the end vertices of a tree T with their neighbors yields the center of T, if the sequence of contraction ends when the resulting graph has either one or two vertices.

3. Find an infinite series of trees whose center is different from the distance center.

4. Show that the distance center of a Cartesian product is the Cartesian product of the distance centers of the factors.

5. Show that the distance center of a graph G is invariant under automorphisms of G.

6. Let f be a contraction of a finite graph G. Show the existence of an integer k that satisfies the property $f^k(G) = f^{k+1}(G)$.

7. Let G, f, and k be defined as in Exercise 6. Show that f restricted to $f^k(G)$ is an automorphism of $f^k(G)$.

8. Let f be a contraction of a graph G. Show the existence of an integer ℓ such that f^ℓ is a retraction of G.

9. Prove Theorem 13.6 with the results of Exercises 8, 7, 6, Theorem 13.5, and Exercise 5.

10. Give an example to show that the result of Exercise 6 need not be true if G is allowed to be infinite.

Canonical Metric Representation

In this chapter, we present one of the most important results about Cartesian products from the area of metric graph theory. It asserts that any graph has a canonical metric representation as an isometric subgraph of a Cartesian product. A graph embeds via the identity mapping into itself, in which case the representation is trivial. However, as soon as the representation gives an isometric embedding into the Cartesian product of more than one factor, the representation yields several useful properties of the embedded graph.

The canonical metric representation has many appealing properties, the most fundamental being that the graph embedded is an isometric subgraph. When we wish to represent a graph as an isometric subgraph of some Cartesian product we would like to have as many factors as possible, provided that none of the factors is useless. In this respect, the canonical representation is unique.

The representation can also be applied in several ways. As a small result, we mention that trees on n vertices can be characterized as those connected graphs for which the canonical metric representation uses $n - 1$ coordinates. At the end of the chapter, we illustrate the usefulness of the representation by the Wiener index, which is very important in chemical graph theory.

14.1 Representation α

We now describe the canonical metric representation of a graph and prove its isometry. First some preparation is needed.

The *Djoković-Winkler relation* Θ is defined [29, 118] on the edge set of a graph G in the following way. Let $xy, uv \in E(G)$, then $xy\Theta uv$ if

$$d(x, u) + d(y, v) \neq d(x, v) + d(y, u).$$

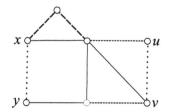

Figure 14.1. Three pairs of edges that are in relation Θ.

Figure 14.1 shows three pairs of edges of a graph that are pairwise in relation Θ. For instance, for the edges xy and uv from the figure we read $d(x,u) = d(y,v) = d(x,v) = 2$, and $d(y,u) = 3$; therefore, $xy\Theta uv$. Several other pairs of edges in addition to the ones marked are in relation Θ (Exercise 1).

Although the definition of Θ seems a little awkward, it is not a difficult relation to use in practice. For practical considerations we collect the following useful facts. (The proofs are left to the reader as Exercise 2.)

Lemma 14.1. *Let G be a graph. Then*

  (i)   *Two adjacent edges of G are in relation Θ if and only if they belong to a common triangle.*

  (ii) *Let P be a shortest path in G. Then no two edges of P are in the relation Θ.*

  (iii) *Let C be an isometric cycle of G and e an edge of C. Then an edge $f \neq e$ of C is in relation Θ with e only when e and f are antipodal edges of C.*

Note that Lemma 14.1(ii) implies that no pair of distinct edges of a tree is in the relation Θ. More generally, edges from different blocks of a (connected) graph will never be in the relation Θ. (See Exercise 3.)

Let C be an isometric even cycle of a graph G. Then Lemma 14.1(iii) implies that an edge e of C is in relation Θ to precisely one edge of C different from e, that is, its unique antipodal edge. If C is an odd isometric cycle, then an edge e of C has two antipodal edges on C.

We continue with another important property of Θ.

Lemma 14.2. *Let $e = uv$ be an edge of a graph G, and let W be a u,v-walk in G that does not contain e. Then there exists an edge f of W such that $e\Theta f$.*

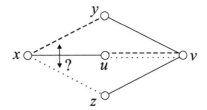

Figure 14.2. The relation Θ is not transitive in $K_{2,3}$.

Proof: Let $W = w_0, w_1, \ldots, w_k$, where $w_0 = u$ and $w_k = v$. Then $k \geq 2$ since e is not on W. Let

$$\lambda = \sum_{i=1}^{k} [d(u, w_{i-1}) + d(v, w_i) - d(u, w_i) - d(v, w_{i-1})]. \qquad (14.1)$$

Expanding the right part of (14.1), we get

$$\lambda = d(u, w_0) - d(v, w_0) + d(v, w_k) - d(u, w_k).$$

Hence, $\lambda = -2$, since $d(u, w_0) = d(v, w_k) = 0$. It follows that at least one of the summands of (14.1) is nonzero. In other words, $e\Theta f$, where $f = w_{i-1}w_i$ for some i, $1 \leq i \leq k$. $\square$

Clearly, Θ is reflexive and symmetric. However, it is not transitive in general, as can be seen on $K_{2,3}$; see Figure 14.2. By Lemma 14.1(iii) (or directly), we infer that $xy\Theta uv$ and $uv\Theta xz$, but xy is not in relation Θ with xz by Lemma 14.1(i).

Let Θ^* denote the transitive closure of Θ. Then Θ^* is an equivalence relation. We shall refer to the equivalence classes of it as Θ^*-*classes*. For instance, it follows easily from Lemma 14.1(i) that a complete graph has a single Θ^*-class. The same holds for $K_{2,3}$ in Figure 14.2.

We are now ready to describe the canonical metric representation α of a graph. Let G be a connected graph, and let $F_1, \ldots, F_k$ be its Θ^*-classes. Then the *quotient graphs* G/F_i, $i = 1, \ldots, k$, are defined as follows: their vertices are the connected components of $G - F_i$, two vertices C and C' being adjacent if there exist vertices $x \in C$ and $y \in C'$ such that $xy \in F_i$. Then α is the mapping

$$\alpha : G \to \square_{i=1}^{k} G/F_i$$

with

$$\alpha : u \mapsto (\alpha_1(u), \ldots, \alpha_k(u)), \qquad (14.2)$$

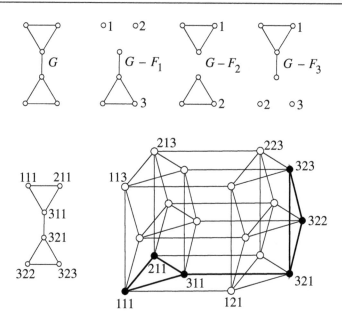

Figure 14.3. Example of the canonical metric representation. Graph G and its corresponding subgraphs (top row), mapping α (lower left), embedding of G into $K_3 \square K_2 \square K_3$ (lower right).

where $\alpha_i(u)$ is the connected component of $G - F_i$ that contains u. The mapping α is called the *canonical metric representation* (of the connected graph G).

Consider the example from Figure 14.3. By Lemma 14.1, it is immediate that G has three Θ^*-classes. The corresponding graphs $G - F_1$, $G - F_2$, and $G - F_3$ are shown in the upper part of the figure. Clearly, $G/F_1 = G/F_3 = K_3$; both factor graphs have vertices $1, 2, 3$; and $G/F_2 = K_2$ with vertices $1, 2$. In the lower part of the figure, the mapping α is indicated on the left side, and the right side shows the embedding of G into $K_3 \square K_2 \square K_3$.

We wish to show that α is an isometry, for which the following lemma is crucial. The proof is omitted because it is rather technical and not used in the sequel. It can be found in [26, 46, 66].

Lemma 14.3. [46] *Let G be a connected graph, F a Θ^*-class of G, and $u, v \in G$. If P is a shortest u, v-path and P' an arbitrary u, v-path, then $|P' \cap F| \geq |P \cap F|$.*

Using Lemma 14.3, Graham and Winkler proved the following important result.

Theorem 14.4. [46] *Let G be a connected graph and α the mapping as defined in (14.2). Then $\alpha(G)$ is an isometric subgraph of $\square_{i=1}^{k} G/F_i$.*

Proof: We show first that α is a bijection from G to $\alpha(G)$. Given two arbitrarily chosen vertices $u, v \in G$, we thus have to show that $\alpha(u) \neq \alpha(v)$. Let P be a shortest u, v-path, and x the vertex of P adjacent to u. (It is possible that $x = v$.) Set $e = ux$, and let P' be any u, v-path that does not contain e. Then the x, v-subpath of P together with P' forms a u, x-walk W that does not contain e. By Lemma 14.2, W contains an edge f with $e \Theta f$, and by Lemma 14.1(ii), $f \notin P$. Hence, if $e \in F_i$, then any u, v-path contains an edge of F_i, which in turn implies that $\alpha_i(u) \neq \alpha_i(v)$. Therefore, $\alpha(u) \neq \alpha(v)$.

Moreover, if u is adjacent to v, then $\alpha_j(u) = \alpha_j(v)$ for any $j \neq i$. Since $\alpha_i(u) \neq \alpha_i(v)$, $\alpha(u)\alpha(v)$ is an edge of $\square_{i=1}^{k} G/F_i$. Thus, α maps edges to edges. In other words, α is a homomorphism.

To prove that $\alpha(G)$ is an isometric subgraph of $H = \square_{i=1}^{k} G/F_i$, we have to show that $d_H(\alpha(u), \alpha(v)) = d_G(u, v)$. Since $\alpha(P)$ is a walk in H, we infer that
$$d_H(\alpha(u), \alpha(v)) \leq d_G(u, v).$$

To complete the proof, it suffices to show that $d_H(\alpha(u), \alpha(v)) \geq d_G(u, v)$ also holds.

Let Q be a shortest $\alpha(u), \alpha(v)$-path in H, i be an arbitrary index from $\{1, \ldots, k\}$, and q_i be the number of edges in Q whose endpoints differ in the ith coordinate. Then there exists a sequence of $q_i + 1$ connected components $X_0, X_1, \ldots, X_{q_i}$ in $G - F_i$, where $u \in X_0$, $v \in X_{q_i}$, such that there is an edge $e_j \in F_i$ between X_{j-1} and X_j for $j = 1, \ldots, q_i$. Then the edges $e_1, \ldots, e_{q_i}$ can be extended by edges from $X_0, X_1, \ldots, X_{q_i}$ to a path R between u and v in G. Since $R \cap F_i = \{e_1, \ldots, e_{q_i}\}$, Lemma 14.3 implies that $q_i \geq |P \cap F_i|$. Therefore,

$$d_H(\alpha(u), \alpha(v)) = |Q| = \sum_{i=1}^{k} q_i \geq \sum_{i=1}^{k} |P \cap F_i| = |P| = d_G(u, v).$$

$\square$

14.2 Properties and Applications of α

The canonical metric representation is a strong tool in the investigation of isometric subgraphs of hypercubes and Cartesian products of complete graphs, which in turn have many applications of their own. For example, it can be shown relatively easily that a given graph is an isometric subgraph of a hypercube or a Cartesian product of complete graphs if and only if the factor graphs G/F_i in the representation

$$\alpha : G \to \square \, G/F_i$$

are either all K_2's or complete; see Exercise 7. It follows that the embedding of such graphs into hypercubes or products of complete graphs are unique in a certain sense and that there are efficient algorithms to find them.

Here we characterize trees in terms of the canonical metric representation, show that it is irredundant, and apply it to the computation of the Wiener index of graphs.

Proposition 14.5. *Let G be a connected graph on n vertices. Then G is a tree if and only if the canonical metric representation of G uses $n - 1$ coordinates.*

Proof: We have already mentioned that every edge of a tree forms a Θ^*-class. Hence, α embeds into a Cartesian product of $n - 1$ factors. (More precisely, it embeds into Q_{n-1}.)

Assume now that G is not a tree. Then it contains a cycle C. By Lemma 14.2, we can find two different edges on C, say f and f', such that $f \Theta f'$. Construct a spanning tree T of G that contains f and f'. (How can one do that? See Exercise 4.) Let e be an arbitrary edge of $G - E(T)$. Then, using Lemma 14.2 again, e is in relation Θ with at least one edge of T. It follows that Θ^* contains at most $n - 2$ classes. □

The canonical representation has many other useful properties in addition to its isometry. Let $\beta : G \to H = \square_{i=1}^t H_i$ be a mapping where $\beta(G)$ is an isometric subgraph of H. Then we say that β is an *irredundant embedding* provided that every factor graph H_i has at least two vertices and that each vertex of H_i appears as a coordinate of $\beta(u)$ for some $u \in G$.

For our next result, the following lemma is useful. (Its proof is left as Exercise 6; it can be proved with the Distance Lemma.)

Lemma 14.6. *Let f and f' be edges of a Cartesian product graph with $f \Theta f'$. Then the end vertices of f and the end vertices of f' differ in the same coordinate.*

Consider the Cartesian product from Figure 14.3. Set $x = 111$, $y = 121$, $u = 321$, and $v = 323$. Then xy is an edge that differs in the second coordinate and uv an edge that differs in the third coordinate. Since $d(x, u) + d(y, v) = 2 + 2 = d(x, v) + d(y, u) = 3 + 1$, we see that xy is not in relation Θ with uv.

Theorem 14.7. [46] *The canonical metric representation of a connected graph G is irredundant. Moreover, it has the largest possible number of factors among all irredundant isometric embeddings and is unique among such embeddings.*

Proof: Let $\alpha : G \rightarrow \Box_{i=1}^{k} G/F_i$ be the canonical metric representation of G. By Lemma 14.2, $G - F_i$ has at least two connected components, so G/F_i has at least two vertices. In addition, the coordinate mappings α_i of α are, by definition, surjective; hence, α is indeed irredundant.

Let $\beta : G \rightarrow H = \Box_{i=1}^{t} H_i$ be an arbitrary isometric irredundant embedding. Since $\beta(G)$ is isometric, Lemma 14.6 implies that no two edges of $\beta(G)$ that lie in different factor fibers are in relation Θ. It follows that α has at least as many factors as β. Moreover, α will have t factors if and only if, for any coordinate i, all the edges of H that differ in the coordinate i are in the same Θ^*-class. But then $\alpha = \beta$. $\Box$

To conclude the chapter, we show that the Wiener index of a graph can be read out from its canonical metric representation.[1]

For Theorem 14.8, it will be convenient to introduce the concept of a weighted graph and the corresponding Wiener index. A *weighted graph* (G, w) is a graph G together with a weight function $w : V(G) \rightarrow \mathbb{R}$. Then the Wiener index of (G, w), $W(G, w)$, is defined as

$$W(G, w) = \frac{1}{2} \sum_{u \in G} \sum_{v \in G} w(u)\, w(v)\, d_G(u, v).$$

If all the weights are 1, then $W(G, w) = W(G)$.

The weights that we use here are defined by the canonical metric representation α. Let G be a graph, and let $\alpha : G \rightarrow \Box_{i=1}^{k} G/F_i$ be its canonical metric representation. To introduce weights for the vertices of the quotient graphs G/F_i, we recall that their vertices are the connected components $C_1^{(i)}, \ldots, C_{r_i}^{(i)}$ of $G - F_i$, $1 \leq i \leq k$. We then set $w_\alpha(C_j^{(i)}) = |C_j^{(i)}|$.

Theorem 14.8. [77] *Let G be a connected graph and let $\alpha : G \rightarrow \Box_{i=1}^{k} G/F_i$ be the canonical metric representation of G. Then*

$$W(G) = \sum_{i=1}^{k} W(G/F_i, w_\alpha).$$

---

[1] Theorem 14.8 is useful only if the investigated graphs have more than one Θ^*-class. This is indeed the case for the graphs of many chemical compounds, for example, benzenoids and phenylenes.

Proof: Set $H = \square_{i=1}^{k} G/F_i$. Then, by Theorem 14.4,

$$W(G) = \frac{1}{2} \sum_{u \in G} \sum_{v \in G} d_G(u, v) = \frac{1}{2} \sum_{u} \sum_{v} d_H(\alpha(u), \alpha(v)).$$

Thus, by the Distance Lemma, $W(G)$ equals

$$\frac{1}{2} \sum_{u} \sum_{v} \sum_{i=1}^{k} d_{G/F_i}(\alpha_i(u), \alpha_i(v)) = \sum_{i=1}^{k} \left(\frac{1}{2} \sum_{u} \sum_{v} d_{G/F_i}(\alpha_i(u), \alpha_i(v)) \right),$$

which is in turn equal to the assertion of the theorem because

$$W(G/F_i, w_\alpha) = \frac{1}{2} \sum_{u} \sum_{v} d_{G/F_i}(\alpha_i(u), \alpha_i(v)).$$

$\square$

Consider the graph G from Figure 14.3. Since all the factor graphs G/F_i are complete, $d_{G/F_i}(C_j^{(i)}, C_{j'}^{(i)}) = 1$ for all indices i, j, j'. Hence,

$$W(G) = (1 \cdot 1 + 1 \cdot 4 + 1 \cdot 4) + (3 \cdot 3) + (4 \cdot 1 + 4 \cdot 1 + 1 \cdot 1) = 27.$$

14.3 Exercises

1. Find all pairs of edges of the graph from Figure 14.1 that are in relation Θ.

2. Prove Lemma 14.1.

3. Let G be a connected graph with distinct blocks B_1 and B_2. Show that if $e_1 \in E(B_1)$ and $e_2 \in E(B_2)$, then e_1 and e_2 are not in the relation Θ.

4. Let G be a connected graph and $f, f' \in E(G)$. Show that there exists a spanning tree of G that contains f and f'.

5. Let e and f be edges of a connected product $G_1 \square \cdots \square G_k$ such that e and f differ in the same coordinate, say i, and that $p_{G_i}(e) = p_{G_i}(f)$. Show that $e \Theta f$.

6. Prove Lemma 14.6.

7. [65] Let G be a connected graph and $\beta : G \to H = \square_{i=1}^{t} H_i$ be an isometric irredundant embedding. Prove that if H_i is a complete graph for every i, then β is the canonical metric representation.

8. [77] Using Theorem 14.8 and the notations above it, show that

$$W(G) \geq \sum_{i=1}^{k} \sum_{1 \leq j < j' \leq r_i} |C_j^{(i)}| \cdot |C_{j'}^{(i)}|,$$

where equality holds if and only if G is an isometric subgraph of the Cartesian product of complete graphs.

9. Design an algorithm for the recognition of isometric subgraphs of products of complete graphs.

10. Let G be a partial cube with the Θ-classes $F_1, \ldots, F_k$. Let $u_i v_i \in F_i$, $1 \leq i \leq k$, be a set of representatives for the Θ-classes of G. Furthermore, for the edge $u_i v_i$, let G_{u_i} be the set of vertices of G that are closer to u_i than to v_i; define G_{v_i} analogously. Show that

$$W(G) = \sum_{i=1}^{k} |G_{u_i}| |G_{v_i}|.$$

Part V

Algebraic and Algorithmic Issues

15 | Prime Factorizations

Despite the fact that many interesting classes of graphs are Cartesian products, almost all graphs are indecomposable. This is easy to show for many cases, such as for trees and complete graphs. For a proof, note that connected graphs have connected factors. If they are nontrivial, every one of them contains an edge. The product of two edges is a square without diagonals; hence, every connected nontrivial Cartesian product contains a square without diagonals. Since trees and complete graphs are connected and contain no squares at all, or at least no squares without diagonals, they are indecomposable, or prime.

This means that hypercubes, Hamming graphs, and integer lattices are products of prime graphs. It is difficult to imagine that any one of these graphs has more than one representation as a product of prime graphs. We would assume these representations to be unique, and indeed, this is the case. In this chapter, we show that all connected graphs have unique prime factor decompositions with respect to the Cartesian product.

We begin with the existence of prime factorizations. Consider a Cartesian product $G = G_1 \square \cdots \square G_k$ of nontrivial factors, that is, of factors with at least two vertices. Then $2^k \leq |G|$ and $k \leq \log_2 |G|$. Thus, every graph G has a decomposition with a largest number of nontrivial factors. Clearly, these factors cannot be representable as a product of nontrivial graphs. In other words, these factors are indecomposable, or *prime*. Therefore, every finite graph G has a prime factorization with respect to the Cartesian product.

The question is whether this factorization is unique. We show in the next section that this is indeed the case for connected graphs.[1] In Section 15.2, we completely describe the automorphism group of a connected graph in terms of the automorphism groups of its prime factors.

---

[1] For disconnected graphs, this need not be true; see Chapter 16.

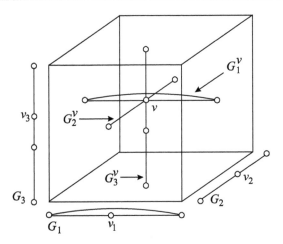

Figure 15.1. Fibers in a product of three factors.

15.1   Uniqueness for Connected Graphs

The uniqueness of the prime factor decomposition of connected graphs
with respect to the Cartesian product was shown by Sabidussi [104],
and independently by Vizing [111].

We present two proofs, one here and the other one in Chapter 18.
The proof in this chapter invokes the convexity of fibers and the fact
that convex subgraphs are boxes. In Chapter 1, we introduced fibers
for products of two factors. Here we extend the definition to products
of several factors. Let $G = G_1 \square \cdots \square G_k$. Then the G_i-*fiber* through
$v \in G$, $v = (v_1, v_2, \ldots, v_k)$, is the subgraph of G induced by the vertex
set

$$\{(v_1, v_2, \ldots, v_{i-1}, x, v_{i+1}, \ldots, v_k) \mid x \in G_i\}.$$

We denote it by G_i^v; see Figure 15.1.

Clearly, every G_i-fiber is isomorphic to G_i. Moreover, Lemma 12.3
easily generalizes to products with more than two factors. Hence, all
the fibers of G are convex.

Theorem 15.1. [104, 111] *Every connected graph has a unique representation as a product of prime graphs, up to isomorphisms and the
order of the factors.*

Since we already know that every finite graph has a prime factorization, we only have to show that it is unique. Therefore, to prove
Theorem 15.1, it suffices to prove Lemma 15.2.

Lemma 15.2. *Let φ be an isomorphism between the connected graphs G and H that are representable as products $G = G_1 \square \cdots \square G_k$ and $H = H_1 \square \cdots \square H_\ell$ of prime graphs, where $k \leq \ell$. Then $k = \ell$, and to every $v \in G$ there exists a permutation π of the set $\{1, 2, \ldots, k\}$ such that $\varphi G_i^v = H_{\pi i}^{\varphi v}$ for $1 \leq i \leq k$.*

Proof: By Lemma 12.3, all fibers are convex. Hence G_i^v and its image φG_i^v are convex, and thus are boxes by Proposition 13.3. Because G_i is prime, all projections $p_j (\varphi G_i^v)$ except one must consist of only one vertex. In other words, φG_i^v must be contained in a fiber of $H_1 \square \cdots \square H_\ell$, say $\varphi G_i^v \subseteq H_{\pi i}^{\varphi v}$, where πi is an index between 1 and ℓ that depends on i, v, and φ.

But then, $G_i^v \subseteq \varphi^{-1} H_{\pi i}^{\varphi v}$. Since $\varphi^{-1} H_{\pi i}^{\varphi v}$ is convex, it is a box, and because it is prime it must coincide with G_i^v. Therefore $G_i^v = \varphi^{-1} H_{\pi i}^{\varphi v}$, and thus $\varphi G_i^v = H_{\pi i}^{\varphi v}$.

Since φ is an isomorphism, every edge incident with φv is in some $\varphi G_i^v = H_{\pi i}^{\varphi v}$, $1 \leq i \leq k$. Hence, $k = \ell$.

Finally, no two distinct G_i^v, G_j^v can have the same image. Thus, π is a permutation of the set $\{1, 2, \ldots, k\}$. $\square$

Theorem 15.1 is an important structure theorem, but it does not tell us how to find the prime factorization of a graph efficiently. For hypercubes or Hamming graphs—that is, the Cartesian products of complete graphs—it is easy to find fast factorization algorithms. But it was not clear for a long time whether the factorization problem could be solved by a polynomial algorithm in general. The first such algorithm was published in 1985 by Feigenbaum, Hershberger, and Schäffer [34] and is of complexity $O(n^{4.5})$, where n is the number of vertices of the graph to be factored. This complexity was subsequently reduced by several authors.

One of these algorithms uses a structure theorem due to Feder [33], which we will also prove as Lemma 18.4 in Chapter 18. It immediately yields a simple algorithm of complexity $O(m^2)$, where m is the number of edges of the graph to be factored, and can be improved to an algorithm of complexity $O(mn)$.

By different methods, Imrich and Peterin [69] derived an algorithm that is linear in the number of edges. The algorithm involves a bucket sort with a bucket for every edge at the start. The buckets are then successively combined. Finally, there is a bucket $b(e)$ for every edge e in a prime factor G_i. The bucket $b(e)$ contains all edges f that are mapped into e by the projection p_i of G onto G_i.

15.2 Automorphisms

The preceding results can be used to describe the structure of the *automorphism group* Aut(G) of connected graphs G in terms of the automorphism groups of the prime factors. First, we prove a corollary to Lemma 15.2.

Corollary 15.3. *Let G, H, φ, v, and π be defined as in Lemma 15.2. Then π depends only on φ and is independent of the choice of v.*

Proof: Let i be a fixed index, $1 \le i \le k$. Since v was arbitrary, the mapping φ maps every G_i-fiber into an H_r-fiber for some r. We wish to show that r depends only on the index i.

Let G_i^u and G_i^w be adjacent G_i-fibers, that is, fibers where at least one vertex of G_i^u is adjacent to a vertex of G_i^w. By the definition of the Cartesian product, these two fibers induce a subgraph isomorphic to $K_2 \square G_i$. Since a single edge of a graph is always convex, as a subgraph this subproduct is convex.

By Lemma 15.2, indices r and s exist such that φG_i^u is an H_r-fiber $H_r^{\varphi u}$, and φG_i^w is an H_s-fiber $H_s^{\varphi w}$. If $r \ne s$, then $H_r^{\varphi u} \cap H_s^{\varphi w}$ is either empty or consists of a single vertex. In both cases, $H_r^{\varphi u} \cup H_s^{\varphi w}$ is not a convex subgraph of H. Thus, $r = s$. Since G is connected, we see by induction with respect to the distance of the fibers that every G_i-fiber must be mapped into an H_r-fiber. $\square$

The next lemma concerns the relationship between the coordinates of $u \in G_1 \square \cdots \square G_k$ and $\varphi u \in G_1 \square \cdots \square G_k$, for an automorphism φ.

Lemma 15.4. *Let $G_1 \square \cdots \square G_k$ be the prime factor decomposition of the connected graph G and $\varphi \in \text{Aut}(G)$. Then there exists a permutation π of $\{1, 2, \ldots, k\}$ with the property that $p_i u = p_i v$ if and only if $p_{\pi i} \varphi u = p_{\pi i} \varphi v$.*

Proof: If we replace $H_1 \square \cdots \square H_k$ by $G_1 \square \cdots \square G_k$ in Corollary 15.3, we see that there is a permutation π of $\{1, 2, \ldots, k\}$ such that every G_i-fiber is mapped into a $G_{\pi i}$-fiber. We now fix i and consider

$$G_i^* = G_1 \square \cdots \square G_{i-1} \square G_{i+1} \square \cdots \square G_k.$$

Then G can be represented as $G_i \square G_i^*$, and the G_i^*-fibers of G are the connected components of the graph that we obtain from G by removing the edges of the G_i-fibers. Recall that u, v are in such a fiber G_i^* if and only if $p_i u = p_i v$.

Since φ maps the edges of the G_i-fibers of G into the edges of the $G_{\pi i}$-fibers of G it is clear that φ maps G_i^*-fibers into the $G_{\pi i}^*$-fibers, where

$$G_{\pi i}^* = G_{\pi 1} \,\square\, \cdots \,\square\, G_{\pi(i-1)} \,\square\, G_{\pi(i+1)} \,\square\, \cdots \,\square\, G_{\pi k}.$$

Hence, $p_{\pi i}\varphi u = p_{\pi i}\varphi v$ if $p_i u = p_i v$. The implication in the other direction is obtained by replacement of φ by φ^{-1}. $\square$

We can now prove the main theorem of this section, Theorem 15.5, which is due to Imrich [62] and Miller [92].

Theorem 15.5. [62, 92] *Let $G = G_1 \,\square\, \cdots \,\square\, G_k$ be the prime factorization of the connected graph G and $\varphi \in \mathrm{Aut}(G)$. Then there exists a permutation π of $\{1, 2, \ldots, k\}$ together with isomorphisms $\psi_i : G_i \to G_{\pi i}$ such that*

$$\varphi(u_1, u_2, \ldots, u_k) = (\psi_{\pi^{-1}1} u_{\pi^{-1}1}, \; \psi_{\pi^{-1}2} u_{\pi^{-1}2}, \; \ldots, \; \psi_{\pi^{-1}k} u_{\pi^{-1}k}).$$

Proof: We arbitrarily choose a vertex v of G and identify, for every index i in $\{1, 2, \ldots, k\}$, the fiber G_i^v with G_i. Then the projection $p_i(u)$ of any vertex u into the ith factor is in G_i^v, and we can summarize the assertion of Lemma 15.4 by the identity

$$\varphi p_i(u) = p_{\pi i}(\varphi u). \tag{15.1}$$

The restriction of φ to G_i^v is an isomorphism from G_i^v to $G_{\pi i}^{\varphi v}$. We denote it by ψ_i. Interchanging sides, Equation (15.1) thus becomes

$$p_{\pi i}(\varphi u) = \psi_i p_i(u). \tag{15.2}$$

But then

$$p_i(\varphi u) = p_{\pi(\pi^{-1}i)}(\varphi u) = \psi_{\pi^{-1}i} \, p_{\pi^{-1}i}(u) = \psi_{\pi^{-1}i} \, u_{\pi^{-1}i},$$

which proves the theorem. $\square$

Note that the conclusion of Theorem 15.5 is equivalent to Equation (15.2). It says that the $(\pi i)^{\mathrm{th}}$ component of the vertex φu is the image of the ith component of u under ψ_i. In other words, map the ith coordinate of u with ψ_i and this becomes the $(\pi i)^{\mathrm{th}}$ coordinate of φu.

To grasp the notation of Theorem 15.5, consider the Cartesian product $G \,\square\, H$, where both factors are isomorphic to P_3. Denote the vertices of the factors by g_1, g_2, g_3 and h_1, h_2, h_3. See Figure 15.2.

Set $\pi = (12)$, and let the mappings $\psi_1 : G \to H$ and $\psi_2 : H \to G$ be defined with

$$\psi_1 g_1 = h_3, \psi_1 g_2 = h_2, \psi_1 g_3 = h_1$$

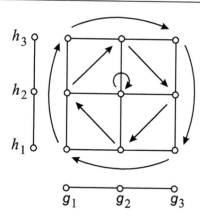

Figure 15.2. An automorphism of $P_3 \,\square\, P_3$.

and

$$\psi_2 h_1 = g_1, \psi_2 h_2 = g_2, \psi_2 h_3 = g_3 \,.$$

Then

$$\varphi(g_i, h_j) = (\psi_2 h_j, \psi_1 g_i), \ 1 \le i, j \le 3 \,.$$

The isomorphism φ is illustrated in Figure 15.2. (For another example, see Exercise 9.)

Let us consider two special cases. If only one ψ_i is nontrivial and π is the identity, then $\varphi(u_1, \dots, u_i, \dots, u_k) = (u_1, \dots, \psi u_i, \dots, u_k)$. We say φ is *generated by an automorphism of a factor*.

If π just interchanges two isomorphic factors of $G = G_1 \,\square\, \cdots \,\square\, G_k$, say the factor G_i by G_j, and all the ψ_i are trivial, then

$$\varphi(u_1, \dots, u_i, \dots, u_j, \dots, u_k) = (u_1, \dots, u_j, \dots, u_i, \dots, u_k).$$

In this case, we speak of a *transposition of isomorphic factors*.

Theorem 15.5 thus tells us that $\mathrm{Aut}(G_1 \,\square\, \cdots \,\square\, G_k)$ is generated by automorphisms of the factors and transpositions of isomorphic factors. This implies the following simple corollary, which helps to visualize the structure of the automorphism group of a product of prime graphs.

Corollary 15.6. *The automorphism group of the Cartesian product of connected, prime graphs is isomorphic to the automorphism group of the disjoint union of the factors.*

Proof: Let $G_1, \dots, G_k$ be the connected components of a graph G. Assume each of these components is a prime graph with respect to Cartesian factorization. Then an automorphism φ of G_i yields an automorphism of G by applying φ on G_i and fixing all vertices of the other components. In addition, if components G_i and G_j are isomorphic, interchanging G_i with G_j and fixing all other vertices also gives an automorphism of G. Every other automorphism of G is generated by automorphisms of these two types. Hence, the structure of the automorphism group of G is the same as that of the automorphism group of the corresponding Cartesian product.   □

15.3 Exercises

1. Given a product $G_1 \square G_2 \square \cdots \square G_k$, show that two fibers G_i^v and G_j^w either are identical or their intersection consists of at most one vertex.

2. Show that the number of vertices in the intersection of a triangle with a fiber of a Cartesian product is zero, one, or three.

3. Suppose u, w are vertices of a Cartesian product with exactly one common neighbor v. Show that uv and vw are in the same fiber.

4. Show that a connected product $G \square H$ of a nontrivial graph G with a graph H on at least three vertices contains two squares that have exactly one edge in common.

5. For which pairs m, n is the complete bipartite graph $K_{m,n}$ prime?

6. Let $u \in G \square H$, and let G^h be an arbitrary fiber of $G \square H$. Show that there exists a unique vertex $v \in G^h$ such that

$$d(u, x) = d(u, v) + d(v, x)$$

for every vertex $x \in G^h$.

7. As in Exercise 6, let $u \in G \square H$, and let G^h be an arbitrary fiber of $G \square H$. Show that the vertex v, whose existence is asserted in Exercise 6, satisfies the condition

$$d(u, v) = \min_{x \in G^h} d(u, x).$$

8. Give an example of a convex subgraph W of a graph G and a vertex $u \in G$ such that there exists no vertex $v \in W$ such that

$$d(u, x) = d(u, v) + d(v, x)$$

for every vertex $x \in W$.

9. Let $G = K_3 \square K_{2,3} \square K_3$. Following the notation of Theorem 15.5, find an explicit automorphism of G that interchanges the first and third factors and acts nontrivially on the second factor.

10. Show that the automorphism group of a product of pairwise non-isomorphic, connected prime graphs is the direct product of the groups of the factors.

11. Let G be a connected prime graph with a trivial automorphism group. Show that the order of $\text{Aut}(G \square G)$ is 2.

12. Show that the order of $\text{Aut}(Q_k)$ is $k!\, 2^k$.

13. Show that $\text{Aut}(Q_k)$ is nonabelian for $k \geq 2$.

14. Let G be a connected prime graph with an abelian automorphism group. Show that $\text{Aut}(G \square G \square G)$ is nonabelian.

15. Let G be a connected prime graph with a nontrivial, abelian automorphism group. Show that $\text{Aut}(G \square G)$ is nonabelian.

16. Show that the automorphism group of the line graph of $K_{m,n}$, where $m \neq n$, is the direct product of the symmetric group S_m by S_n.

16

Cancelation and Containment

Let us now return to the beginning of the book and reconsider the Cartesian product operation. We have noticed in Chapter 1 that Cartesian multiplication is

- commutative: $G \square H = H \square G$,
- associative: $(G \square H) \square K = G \square (H \square K)$, and that
- the one vertex graph K_1 is a unit.

We will write $G + H$ to denote the *disjoint union* of G and H, that is, the graph with $V(G + H) = V(G) \cup V(H)$ and $E(G + H) = E(G) \cup E(H)$. Then it is not difficult to see (Exercise 1) that Cartesian multiplication is left- and right-distributive with respect to the disjoint union:

- $G \square (H + K) = G \square H + G \square K$ and
- $(G + H) \square K = G \square K + H \square K$.

Thus, graphs form a commutative semiring with respect to Cartesian multiplication and disjoint union. Note that the empty graph is the neutral element with respect to addition, and K_1 is the identity with respect to multiplication in this semiring (of simple graphs Γ). We denote the semiring by $\mathbb{S}(\Gamma)$.

To obtain additional algebraic properties of the Cartesian multiplication the unique prime factorization of connected graphs (Theorem 15.1) turns out to be a very useful tool. For connected graphs, it immediately implies the cancelation property and the uniqueness of rth roots; that is, $G^r = H^r$ implies $G = H$, where G^r denotes the r-fold Cartesian product of G. We show this in the next section of this chapter.

For disconnected graphs, the situation is more subtle. In this larger class, we no longer have unique prime factorization, but we still have cancelation and uniqueness of rth roots. We show this by embedding

$\mathbb{S}(\Gamma)$ into a polynomial ring with the unique prime factorization property.

For positive integers m, n we know that $m^r \le n^r$ implies $m \le n$. We might thus be tempted to investigate whether the containment $G^r \subseteq H^r$ implies $G \subseteq H$. This is treated in the second section of this chapter.

16.1   Cancelation Properties

Let G and H be connected, nonisomorphic graphs. Let $G = \Box_{i=1}^{n} G_i$ and $H = \Box_{i=1}^{m} H_i$ be the unique prime factorizations of G and H guaranteed by Theorem 15.1. Since G and H are not isomorphic, their factorizations are different. If G has a prime factor P that is not a factor of H, then P is a factor of G^r but not of H^r, and thus $G^r \ne H^r$. If G and H have the same prime factors, there must be one, say P, with different multiplicities in G and H. But then P also has different multiplicities in the rth powers of G and H, and $G^r \ne H^r$. In other words, the following implication

$$G^r = H^r \Rightarrow G = H \tag{16.1}$$

holds for any connected graphs G and H and any positive integer r. Similarly, we can also show (Exercise 2) that the *cancelation law*

$$G \Box K = H \Box K \Rightarrow G = H \tag{16.2}$$

holds for any connected graphs G, H, and K.

The main result of this section asserts that the two implications (16.1) and (16.2) hold for arbitrary graphs, that is, for disconnected as well as connected graphs. To establish these results, we follow the approach of Fernández, Leighton, and López-Presa [35].

Clearly, the number of (prime) graphs of a given order is finite. Therefore, the enumerable set of connected, prime graphs can be listed as $Y_1, Y_2, \ldots$. Let $X = \{x_1, x_2, \ldots\}$ be a denumerable set of variables and $\mathbb{Z}[X]$ the ring of polynomials in these variables with integer coefficients. It is well known that the ring $\mathbb{Z}[X]$ is a unique factorization domain and that it has no zero divisors (see, e.g., Lang [87]). Therefore, $\mathbb{Z}[X]$ satisfies the cancelation property with respect to multiplication. That is, if $f, g, h \in \mathbb{Z}[X]$ such that $f \ne 0$ and $fg = fh$, then $g = h$.

Define a one-to-one correspondence $Y_i \leftrightarrow x_i$ where the empty graph corresponds to $0 \in \mathbb{Z}[X]$ and K_1 corresponds to $1 \in \mathbb{Z}[X]$. In this way we obtain a one-to-one homomorphism of the semiring $\mathbb{S}(\Gamma)$ of all (connected or disconnected) finite graphs into $\mathbb{Z}[X]$, where Cartesian multiplication corresponds to the multiplication of $\mathbb{Z}[X]$ and disjoint union corresponds to the addition of $\mathbb{Z}[X]$. (The latter correspondence motivated our notation for the disjoint union of graphs.)

Every graph G is thus uniquely representable in $\mathbb{Z}[X]$ as a polynomial with positive coefficients. We denote this polynomial by $P(G)$. Then for any graphs G and H,

$$P(G \square H) = P(G) P(H) \tag{16.3}$$

and

$$P(G + H) = P(G) + P(H). \tag{16.4}$$

(The reader is asked to verify Equations (16.3) and (16.4) in Exercise 3.)

In this way, Fernández, Leighton, and López-Presa [35] proved the uniqueness of rth roots, and Imrich, Klavžar, and Rall [68] proved the cancelation property for disconnected graphs with respect to Cartesian multiplication. We begin with the easier result.

Theorem 16.1. [68] *Let G, H, and K be arbitrary graphs. If $G \square K = H \square K$, then $G = H$.*

Proof: Since $G \square K = H \square K$, we have $P(G \square K) = P(H \square K)$, and therefore, using (16.3), $P(G)P(K) = P(H)P(K)$. Since none of $P(G), P(H)$, or $P(K)$ is the zero polynomial, and since $\mathbb{Z}[X]$ satisfies the cancelation property, $P(G)P(K) = P(H)P(K)$ implies $P(G) = P(H)$. We conclude that $G = H$. $\qquad\square$

Theorem 16.2. [35] *Let G and H be arbitrary graphs. If $G^r = H^r$ holds for a positive integer r, then $G = H$.*

Proof: The polynomial $P(G)$ has a unique prime factorization in the ring $\mathbb{Z}[X]$. Let $p_1, p_2, \ldots p_n$ denote all of these prime polynomials with repetitions allowed. That is, $P(G) = p_1 \cdot p_2 \cdots p_n$. Since $P(G^r) = (P(G))^r$, we get

$$P(G^r) = p_1^r \cdot p_2^r \cdots p_n^r. \tag{16.5}$$

But $P(G^r)$ has a unique factorization in $\mathbb{Z}[X]$ as a product of primes, and hence this factorization is the one given in Equation (16.5).

In an entirely similar way, there are prime polynomials $q_1, q_2, \ldots, q_m$ in $\mathbb{Z}[X]$ such that $P(H) = q_1 \cdot q_2 \cdots q_m$ and

$$P(H^r) = q_1^r \cdot q_2^r \cdots q_m^r. \tag{16.6}$$

Since $G^r = H^r$, Equations (16.5) and (16.6) imply that

$$p_1^r \cdot p_2^r \cdots p_n^r = q_1^r \cdot q_2^r \cdots q_m^r.$$

The fact that p_i and q_j are prime in the ring $\mathbb{Z}[X]$ for each $1 \le i \le n$ and each $1 \le j \le m$ now leads immediately to the conclusion that $n = m$ and that

$$p_1 \cdot p_2 \cdots p_n = q_1 \cdot q_2 \cdots q_n.$$

Since the semiring homomorphism, P, is one-to-one, we conclude that $G = H$.
□

We wish to point out a subtlety in the proof of Theorem 16.2. $P(G^r) = P(H^r)$ implies $P(G) = P(H)$ because the factorization in $\mathbb{Z}[X]$ is unique. In general, some of the prime polynomials $p_1, p_2, \ldots, p_n$ in the factorization of $P(G)$ could contain negative coefficients, in which case the corresponding factor does not represent a graph. Nevertheless, this fact is not an obstacle for our proof because in the end we require only that their product, $p_1 p_2 \cdots p_n$, has nonnegative coefficients.

On the other hand, in the set of polynomials with nonnegative coefficients over $X = \{x_1, x_2, \ldots\}$, the unique factorization does not hold. This observation from 1950 is due to Nakayama and Hashimoto [95]. For instance,

$$(1 + x_1 + x_1^2)(1 + x_1^3) = (1 + x_1^2 + x_1^4)(1 + x_1) \qquad (16.7)$$

are different factorizations of the polynomial $x_1^5 + x_1^4 + x_1^3 + x_1^2 + x_1 + 1$. This leads to the following result:

Theorem 16.3. *Prime factorization is not unique in the class of disconnected graphs.*

Proof: Recall that K_1 is assigned to the polynomial 1. Assigning K_2 to the indeterminant x_1, the polynomial factorizations (16.7) yield the following graph factorizations

$$(K_1 + K_2 + K_2^2) \,\square\, (K_1 + K_2^3) = (K_1 + K_2^2 + K_2^4) \,\square\, (K_1 + K_2), \qquad (16.8)$$

of the graph $K_1 + K_2 + K_2^2 + K_2^3 + K_2^4 + K_2^5$. To complete the proof, one needs to verify that the factors in (16.8) are prime (Exercise 4).
□

16.2   Containment Properties

The results of the previous section lead to the question of whether the containment $G^r \subseteq H^r$ (for some $r > 1$) implies that $G \subseteq H$. We are going to show that this need not be the case. For this purpose, consider the following example:

$$G = K_3 \,\square\, P_4 \quad \text{and} \quad H = (K_3 \,\square\, K_3) + (P_4 \,\square\, P_4). \qquad (16.9)$$

Then clearly $G \nsubseteq H$. However, if

$$G^2 = K_3^2 \,\square\, P_4^2$$

and

$$H^2 = K_3^4 + 2(K_3^2 \,\square\, P_4^2) + P_4^4,$$

then $G^2 \subseteq H^2$.

Note that here H is not connected. It is not difficult to modify the construction in such a way that both G and H are connected but still $G^2 \subseteq H^2$ does not imply $G \subseteq H$ (see Exercise 7).

The above examples are due to Fernández, Leighton, and López-Presa [35]. They go further by showing that even if G and H are connected and *prime*, the conclusion still does not hold (Exercise 8). Finally, by an exhaustive computer search, the authors were also able to find two connected graphs G and H of the *same order* with $G \nsubseteq H$ but $G^2 \subseteq H^2$.

16.3 Exercises

1. Show for any graphs G, H, and K, that $G \,\square\, (H+K) = G \,\square\, H + G \,\square\, K$ and $(G + H) \,\square\, K = G \,\square\, K + H \,\square\, K$.

2. Show for any connected graphs G, H, and K, that $G \,\square\, K = H \,\square\, K$ implies $G = H$.

3. Verify Equations (16.3) and (16.4).

4. Show that $(K_1 + K_2 + K_2^2) \,\square\, (K_1 + K_2^3)$ and $(K_1 + K_2^2 + K_2^4) \,\square\, (K_1 + K_2)$ are two distinct prime factorizations of one and the same graph.

5. Use polynomial factorization in $\mathbb{Z}[X]$ to help find a factorization of the disconnected graph

$$C_4 + K_2^2 \,\square\, K_3 + 2(K_2 \,\square\, K_3) + K_2 \,\square\, K_3^2 + K_3^2$$

   into prime factors. Show that the prime factorization is unique.

6. Prove that $C_4^2 \nsubseteq C_6^2$.

7. Modify the graph H from (16.9) such that it becomes connected and such that $G \nsubseteq H$ and $G^2 \subseteq H^2$.

8. Show that there exist connected, prime graphs G and H such that $G \not\subseteq H$ and $G^2 \subseteq H^2$.

9. Show that a disconnected, vertex-transitive graph has a unique prime factorization.

17 Distinguishing Number

Most graphs are asymmetric; that is, the identity mapping is their only automorphism. We could also call them rigid. Well-known graphs such as hypercubes, complete graphs, complete bipartite graphs, the Petersen graph, stars, and lattices admit nontrivial automorphisms. Some other classes, such as Cayley graphs, are even defined via group actions. To understand them, it is essential to understand their symmetries. Sometimes we may even look for methods to break them. Who does not know the problem of where to place nails in order to stop a wooden structure from shifting in undesirable ways? It helps to have insight into the structure to solve this problem intelligently.

This leads to the introduction of the distinguishing number. Given a graph G, the *distinguishing number* $D(G)$ is the least natural number d such that G has a labeling with d labels that is preserved only by the trivial automorphism. This definition is due to Albertson and Collins [2] and has opened a rich, new field of research.

Since we just determined the automorphism groups of Cartesian products, in this chapter we use the insight gained to investigate the distinguishing numbers of Cartesian products. We begin with several basic properties of the distinguishing number, present upper and lower bounds, and determine the distinguishing number for Cartesian powers of graphs.

17.1 Examples and Products of Two Factors

It is easily seen that $D(K_n) = n$ and that every asymmetric graph has distinguishing number 1. Also, $D(P_n) = 2$ for $n \geq 2$, and $D(C_n) = 2$ for $n \geq 6$; compare Figure 17.1. For $n = 3, 4, 5$ we have $D(C_n) = 3$. We leave the proof to the reader. (See Exercise 1.)

If one considers P_n as a subgraph of C_n, then every automorphism of P_n is induced by one of C_n. In such a case, the distinguishing num-

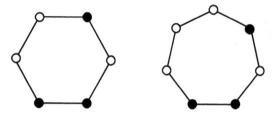

Figure 17.1. Distinguishing C_6 (left) and C_7 (right) with two labels.

ber of the subgraph is bounded by that of the bigger graph. More generally, if G, H have the same vertex sets and if $\text{Aut}(H) \subseteq \text{Aut}(G)$, then $D(H) \leq D(G)$. In particular, we always have $D(G) \leq D(K_{|G|})$.

From the examples, it is clear that we primarily need knowledge of the automorphism group of a graph as a permutation group acting on the set of vertices. This group is identical to that of the complement of the graph. Thus, a graph and its complement have the same distinguishing number, and we restrict our attention here to connected graphs.

For most graphs, the largest degree is an upper bound for the distinguishing number. Consider a connected graph G that is not regular. It contains a vertex u with $\deg(u) < \Delta = \Delta(G)$. We assign the label Δ to it, but to no other vertex. Thus, u is fixed by every label-preserving automorphism of G. Now we arrange the vertices of G in a breadth-first search order with the root u, and proceed as follows: Let v be a vertex in this order. We label its still unlabeled neighbors with different labels from $\{1, 2, \ldots, \Delta - 1\}$ and proceed to the next vertex. The result is a distinguishing labeling of G with Δ labels.

This need not be true if G is regular. To see this, recall that $D(K_n) = n$ and that $D(C_5) = 3$. Moreover, $D(K_{n,n}) = n + 1$ (see Exercise 9). As it turns out, these examples are the only exceptions. The proof can be led with similar, but slightly more sophisticated techniques than the ones used in the regular case; see Collins and Trenk [24] and Klavžar, Wong, and Zhu [81]. In conclusion, the following theorem holds. It is similar to the classic theorem of Brooks [18], Theorem 8.5.

Theorem 17.1. [24, 81] *Let G be a connected graph. Then $D(G) \leq \Delta(G)$ unless G is either K_n, $n \geq 1$, $K_{n,n}$, $n > 1$, or C_5. In these cases, $D(G) = \Delta(G) + 1$.*

Note that the obvious exceptions, C_3 and C_4, appear as K_3 and $K_{2,2}$ in the statement of Theorem 17.1.

We continue with bounds for the distinguishing number of products of relatively prime graphs, where graphs G and H are called *rel-*

atively prime if they share no common factor in their prime factorizations. We see that the distinguishing number of graphs that have approximately the same order is small, but that it can become arbitrarily large for graphs of different order.

Proposition 17.2. *Let $k \geq 2$, $d \geq 2$, G be a connected graph on k vertices, and H be a connected graph on $d^k - k + 1$ vertices that is relatively prime to G. Then $D(G \square H) \leq d$.*

Proof: Since G and H are relatively prime, every automorphism maps G-fibers into G-fibers and H-fibers into H-fibers. Refer to Equation (15.2).

Denote the set of vectors of length k with integer entries between 1 and d by $\mathbb{N}_d^k$, and let S be the set of the following $k - 1$ vectors from $\mathbb{N}_d^k$:

$$(1, 1, 1, \ldots, 1, 1, 1, 2)$$
$$(1, 1, 1, \ldots, 1, 1, 2, 2)$$
$$(1, 1, 1, \ldots, 1, 2, 2, 2)$$
$$\vdots$$
$$(1, 2, 2, \ldots, 2, 2, 2, 2).$$

Consider the $d^k - k + 1$ vectors from $\mathbb{N}_d^k - S$ and label the G-fibers with them. Select an arbitrary coordinate of the vectors in $\mathbb{N}_d^k$. Then there are precisely $d^k / d = d^{k-1}$ vectors from $\mathbb{N}_d^k$ that have the selected coordinate equal to 1. Thus, the number of 1's in the H-fibers is $d^{k-1} - k + 1, \ldots, d^{k-1} - 1, d^{k-1}$, respectively. Hence, any label preserving automorphism φ of $G \square H$ preserves these fibers individually, so φ can permute only the G-fibers. But since they are all different, it follows that φ is the identity. Therefore, the described labeling is d-distinguishing. □

Proposition 17.3. *Let $k, d \geq 2$ and $n > d^k$. Then $D(K_k \square K_n) \geq d + 1$.*

Proof: Let ℓ be an arbitrary d-labeling of $K_k \square K_n$. Since the number of K_k-fibers is more than d^k, at least two of them have identical labels. Furthermore, since $\text{Aut}(K_k \square K_n)$ acts transitively on the K_k-fibers, we infer that ℓ is not distinguishing. Hence $D(K_k \square K_n) \geq d + 1$. □

A more general result for the distinguishing number of the Cartesian product of two complete graphs is listed below. In most cases, it allows us to compute $D(K_k \square K_n)$ via a simple formula, but sometimes a recursion is required. This theorem is due to Fisher and Isaak [38] (who investigated distinguishing edge colorings of the complete bipartite graph $K_{k,n}$) and independently to Imrich, Jerebic, and Klavžar [64].

Theorem 17.4. [38, 64] *Let k, n, d be integers so that $d \geq 2$, $k < n$, and $(d-1)^k < n \leq d^k$. Then*

$$D(K_k \,\square\, K_n) = \begin{cases} d, & \text{if } n \leq d^k - \lceil \log_d k \rceil - 1; \\ d+1, & \text{if } n \geq d^k - \lceil \log_d k \rceil + 1. \end{cases}$$

If $n = d^k - \lceil \log_d k \rceil$, then $D(K_k \,\square\, K_n)$ is either d or $d+1$. It can be computed recursively in $O(\log^* n)$ time, where $\log^*$ denotes the iterated logarithm.

From the book of Cormen, Leiserson, Rivest, and Stein [25], we infer that the iterated logarithm of 2^{65536} is 5, and that the number of atoms in the observable universe is far smaller than 2^{65536}. Hence, the number of iterations needed for the computation of $D(K_k \,\square\, K_n)$ will rarely be larger than five. Nonetheless, $\log^* n$ tends to infinity with n, and we wonder whether one can construct numbers k_i, n_i for every i such that the number of iterations needed for the computation of $D(K_{k_i} \,\square\, K_{n_i})$ is at least i.

17.2   Cartesian Powers of Graphs

We now consider powers of graphs with respect to the product. To simplify the notation we write G^k to denote the k-fold Cartesian product of G. That is, $G^k = \square_{i=1}^k G$.

The distinguishing number is small for products of graphs of similar order. Thus, it comes as no surprise that, with two exceptions, the distinguishing number of the second and all higher powers of a nontrivial connected graph is two. The exceptions are K_2 and K_3. For these, we have $D(K_2^k) = 2$ if $k \geq 4$, and $D(K_3^k) = 2$ if $k \geq 3$.

Proposition 17.5. *K_n^k contains an induced path S with the property that the identity automorphism is the only automorphism of K_n^k that fixes S pointwise.*

Proof: Let $G = G_1 \,\square\, G_2 \,\square\, \cdots \,\square\, G_k$, where every G_i is a K_n and $V(K_n) = \{1, 2, \ldots, n\}$. Then we define S on the vertices

$$\bigcup_{j=1}^{n-1} \{(j, j, \ldots, j), (j+1, j, \ldots, j), \ldots, (j+1, j+1, \ldots, j+1)\},$$

and arrange them in lexicographic order. Thus, the first vertex is $(1, 1, \ldots, 1)$ and the last is $(n, n, \ldots, n)$. It is easy to see that S is an induced path. We leave the details to the reader as Exercise 3.

Let φ be an automorphism of G that fixes S pointwise. By Theorem 15.5, and since K_n is prime, there exist a permutation π of $\{1, 2, \ldots, k\}$ and isomorphisms $\psi_i : G_i \to G_{\pi i}$ such that

$$p_{\pi i}(\varphi u) = \psi_i p_i(u).$$

Consider two adjacent vertices $u = (j+1, \ldots, j+1, j, j, \ldots, j)$ and $v = (j+1, \ldots, j+1, j+1, j, \ldots, j)$ of S that differ in the ith coordinate. Then

$$u_{\pi i} = p_{\pi i}(u) = p_{\pi i}(\varphi u) = \psi_i p_i(u) = \psi_i j \neq$$
$$\psi_i(j+1) = \psi_i p_i(v) = p_{\pi i} \varphi v = v_{\pi i}.$$

Thus, $u_{\pi i} \neq v_{\pi i}$, and therefore $\pi i = i$, because u and v differ only in the ith coordinate. But then $j = u_{\pi i} = \psi_i j$.

Since i and j were arbitrarily chosen, we infer that π and all ψ_i are the identity. Therefore φ is the identity, too.   □

Proposition 17.6. *Let G be a connected prime graph on $n \geq 2$ vertices and let $k \geq 2$ be an integer such that $n + k \geq 6$. Then $D(G^k) = 2$.*

Proof: Consider G as a spanning subgraph of K_n. Clearly $\text{Aut}(G) \subseteq \text{Aut}(K_n)$, and, by Theorem 15.5, $\text{Aut}(G^k) \subseteq \text{Aut}(K_n^k)$. Therefore, every distinguishing labeling of K_n^k also distinguishes G^k. It thus suffices to prove the proposition for $G = K_n$.

- **Case 1:** $k \geq 4$. Add the vertex $u = (1, 1, 2, 1, \ldots, 1)$ to the path S of Proposition 17.5. It is adjacent to $v = (1, 1, \ldots, 1)$ but to no other vertex of S. Therefore $S' = S \cup \{u\}$ is an induced path, too. The endpoint u has distance two from the vertices $(2, 1, 1, 1, \ldots, 1)$ and $(2, 2, 2, 1, \ldots, 1)$ of S, whereas to the other endpoint $(n, n, \ldots, n)$ there exists only one vertex of distance two in S', namely, $(n, n, \ldots, n, n-1, n-1)$. Thus, every automorphism of K_n^k that stabilizes S' fixes every vertex of it.

  If we label the vertices of S' black and all the others white, we obtain a distinguishing 2-labeling of K_n^k by Proposition 17.5.

- **Case 2:** $k = 3$. Then $n \geq 3$. We set $u = (1, 3, 1)$, consider $S' = S \cup \{u\}$, and label the vertices of S' black and the others white. As before, S' is an induced path. Since there are at least four vertices at distance two from u in S', but at most two (only one if $n \geq 4$) at distance two from (n, n, n), we infer that every automorphism of K_n^3 that stabilizes S' fixes every vertex of it. Hence, our labeling is distinguishing.

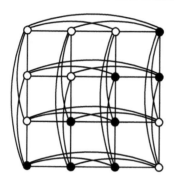 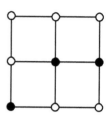

Figure 17.2. Two distinguishing labelings.

- **Case 3:** $k = 2$. Then $n \geq 4$ and we consider $S' = S \cup (3, 1)$. Again, we label the vertices of S' black and the others white.

The vertex $u = (3, 1)$ is fixed because it is the only black vertex of degree four in S'. Furthermore, $v = (1, 1)$ is its only neighbor of degree two, and thus it is also fixed. The vertex $w = (2, 1)$ is the only common neighbor of the two fixed vertices v and u, and thus w is fixed as well. The remaining vertices of S' are on a path from x to (n, n) and thus are also fixed; see Figure 17.2 (left).

By Proposition 17.5, this is a distinguishing 2-labeling.   □

Now we can prove the main result of this chapter, which is due to Imrich and Klavžar [67].

Theorem 17.7. [67] $D(G^k) = 2$ for $k \geq 2$ for all nontrivial, connected graphs $G \neq K_2, K_3$. Furthermore, $D(K_n^k) = 2$ for $n = 2, 3$, if $n + k \geq 6$.

Proof: Proposition 17.6 covers the case when G is prime, with the sole exception of $G = P_3$, where we still need a distinguishing 2-labeling of P_3^2. The right side of Figure 17.2 exhibits such a labeling. (See Exercise 2.)

If G is not prime, let $G_1, G_2, \ldots, G_r$, $r \geq 2$, be the different prime factors of G. Then G^k can be represented in the form

$$G^k = G_1^{s_1} \,\square\, G_2^{s_2} \cdots \,\square\, G_r^{s_r},$$

where every s_i is a multiple of k. The automorphisms of G^k are induced by automorphisms of the $G_i^{s_i}$. We now replace each G_i by a complete graph on $|G_i|$ vertices, set $H_i = K_{|G_i|}^{s_i}$, and consider

$$H = H_1 \,\square\, H_2 \,\square\, \cdots \,\square\, H_r.$$

It is possible that some of the G_i have the same number of vertices, although they are nonisomorphic prime graphs. We thus restrict our attention to the subgroup A of the automorphism group of H that is induced by the groups of the H_i. It contains the automorphism group of G^k. All we need now to complete the proof is a 2-labeling of H such that every $\varphi \in A$ that preserves it is the identity.

We thus choose a vertex v in H and consider the fibers H_i^v. Every such fiber is the power of a complete graph. We select a path S_i in it with the properties of Proposition 17.5. Without loss of generality, we can assume that v is an endpoint of every S_i. We now label all vertices in $T = \cup_{i=1}^r S_i$ black and the others white. Every automorphism φ of H that preserves this labeling must fix v, and if $\varphi \in A$, also every vertex of each S_i. But then φ is the identity on every fiber H_i^v by Proposition 17.5, and thus the identity automorphism of H by Theorem 15.5.   □

17.3 Exercises

1. Determine $D(C_n)$ for $n = 3, 4, 5$.

2. Prove that the 2-labeling of P_3^2 on the right side of Figure 17.2 is distinguishing.

3. Prove that the path S in the proof of Proposition 17.5 is induced.

4. Prove that $K_3 \,\square\, K_3$ and Q_3 have distinguishing 3-labelings.

5. Prove that $D(K_3 \,\square\, K_3) = 3$.

6. Prove that $D(Q_3) = 3$.

7. Let G be an asymmetric, nontrivial graph and K_n a complete graph on at least two vertices. Show that G and K_n are relatively prime.

8. Let G be an asymmetric graph of order k. Prove that $D(G \,\square\, K_n) = \lceil n^{\frac{1}{k}} \rceil$.

9. Prove that $D(K_{n,n}) = n + 1$.

10. [38] Determine the minimal number $\ell(m, n)$ of colors needed to color the edges of the complete bipartite graph $K_{m,n}$ such that the identity automorphism of $K_{m,n}$ is the only automorphism that preserves the colors.

11. [81] A connected graph G is called *uniquely distinguishable* if, for any $D(G)$-distinguishing labelings ℓ_1 and ℓ_2 of G, there exists an automorphism φ of G such that for any vertex $x \in G$ we have $\ell_1(x) = \ell_2(\varphi(x))$. Show that $D(G + G) \le D(G) + 1$ and that the equality holds if and only if G is uniquely distinguishable. (Recall that the operation + stands for the disjoint union of graphs.)

12. Show that C_6 and $K_2 \mathbin{\square} K_3$ are uniquely 2-distinguishable graphs.

Recognizing Products and Partial Cubes

In Chapter 15, we proved the unique prime factorization property of connected graphs with respect to the Cartesian product. The proof did not provide us with a fast, or at least polynomial, algorithm to find the factorization. Here we exploit the close relationship between the relation Θ and the Cartesian product to prove a structure theorem that has both the unique prime factorization property of connected graphs and a polynomial factorization algorithm as immediate consequences.

Then we continue with an example of the canonical isometric embedding. In this example, we illuminate the structure of partial cubes—that is, isometric subgraphs of hypercubes—and derive a recognition algorithm.

18.1 The Structure Theorem

Let G be a Cartesian product $\square_{i=1}^{k} G_i$, where the factors are not necessarily prime. With respect to this representation, we introduce a *product relation* σ on $E(\square_{i=1}^{k} G_i)$ as follows:

$$e\,\sigma\,f \quad \text{if there is an } i \text{ such that} \quad |p_i e| = |p_i f| = 2.$$

Clearly, $|p_j e| = |p_j f| = 1$ for all $j \neq i$ in this case. Besides, the relation σ is transitive, reflexive, and symmetric. By Lemma 14.6, $\Theta \subseteq \sigma$, and since σ is transitive, we infer that $\Theta^* \subseteq \sigma$ also holds. Hence, σ-classes are unions of Θ^*-classes.

In general Θ^* is not a product relation itself. We wish to find out which Θ^*-classes one has to combine to obtain a product relation. This will be accomplished by the following new relation on the edge set of G.

The edges $e = uv$ and $f = uw$ are in the *relation* τ, if the vertex u is the only common neighbor of the nonadjacent vertices v and w. See Figure 18.1.

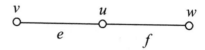

Figure 18.1. Edges e and f in relation τ.

Note that edges that are in the relation τ do not belong to a chord-less square. Therefore, they must be in the same equivalence class with respect to any product relation σ. But then $(\Theta \cup \tau)^* \subseteq \sigma$ for any σ, because σ is transitive. We set $\Pi = (\Theta \cup \tau)^*$ and say that Π satisfies the Square Property provided that any pair of adjacent edges from different Π-classes belong to a unique chordless square.

Lemma 18.1. *The relation Π is contained in every product relation and satisfies the Square Property.*

Proof: We have already shown the first assertion. For the second assertion, suppose that $e = uv$ and $f = uw$ belong to different Π-classes. If uvw is a triangle, then $e\Theta f$ and $e\Pi f$. Similarly, $e\tau f$ implies $e\Pi f$.

The only possibility left is that there are at least two squares without triangles that contain e and f, say $vuwx$ and $vuwy$. Then v, w together with u, x, y form a $K_{2,3}$. But in a $K_{2,3}$, any two edges are in the relation Θ^*, and thus $e\Pi f$. (Compare Figure 14.2.)   □

Theorem 18.2 (Structure Theorem). [33] Π *is a product relation.*

Proof: Let G be a connected graph with the Π-classes $E_1, \ldots, E_k$. Then the *quotient graphs* G/E_i, $i = 1, \ldots, k$, are defined with respect to the E_i just as the G/F_i were defined in Chapter 14. The vertices are the connected components of $G - E_i$, two vertices C and C' being adjacent if there exist vertices $x \in C$ and $y \in C'$ such that $xy \in E_i$. Let φ be the mapping

$$\varphi : G \to \square_{i=1}^{k} G/E_i$$

with

$$\varphi : u \mapsto (\varphi_1(u), \ldots, \varphi_k(u)),$$

where $\varphi_i(u)$ is the connected component of $G - E_i$ that contains u. We wish to show that φ is an isomorphism.

As in the proof of Theorem 14.4, we show that φ preserves edges and that it is injective.

Hence, $\varphi(G)$ is a connected subgraph of $\square_{i=1}^{k} G/E_i$. By definition, every φ_i is surjective. To complete the proof, it suffices to show that $\varphi(G)$ is a box. By Proposition 13.2, this is true because Π satisfies the Square Property.   □

Note that Lemma 14.3, whose proof we omitted because it is rather technical, was not invoked in the above considerations.

We conclude this section with another proof of the unique prime factorization property of connected graphs.

Theorem 18.3. [104, 111] *Every connected graph G has a unique prime factor decomposition with respect to the Cartesian product.*

Proof: From Theorem 18.2, we know that Π is a product relation. Suppose it does not correspond to a prime factorization. Then there must be a $\sigma \subseteq \Pi$ that does. Since Π is contained in any product relation, this implies

$$\Pi \subseteq \sigma \subseteq \Pi.$$

$\square$

18.2 A Polynomial Factorization Algorithm

In this section, we derive a polynomial factorization algorithm based on the Structure Theorem. We wish to compute $\Pi = (\Theta \cup \tau)^*$, the transitive closure of a relation. If we know all elements in a relation ρ, then the computation of the transitive closure amounts to the computation of a connected component in a graph and is thus not more complex than the number of elements in ρ. This can be large if ρ has many elements. Consider $K_2 \,\square\, P_n$, for example. It has a Θ-class that consists of $\binom{n}{2} = O(m^2)$ pairs of edges, where $m = |E(K_2 \,\square\, P_n)| = O(n)$. On this basis, we cannot expect the complexity for the computation of Π to be better than $O(m^2)$.

What we really need, though, are the equivalence classes of Π—that is, a partition of the set $E(G)$ that has cardinality m.

We begin with an elegant, fast way to compute Θ^* that is due to Feder [33]. He introduced a relation Θ_1 that has the same transitive closure as Θ, but fewer elements. We state his result in the following proposition. The proof is technical, similar to the proof of Lemma 14.3, and omitted here.

Lemma 18.4. [33] *Let G be a connected graph, T be a spanning tree, and Θ_1 be the relation that consists of all pairs of edges e, f, with $e \Theta f$ and $\{e, f\} \cap E(T) \neq \varnothing$. Then $\Theta_1^* = \Theta^*$.*

Proposition 18.5. [33] *Let G be a connected graph with m edges and n vertices. Then Θ^* can be computed in $O(mn)$ time.*

Proof: The definition of Θ_1 involves distances. We thus compute the distance matrix first. It is well known that this can be done in $O(mn)$ time.

For every pair uv, xy of edges in $E(T) \times E(G)$, we now look up the distances between the vertices, perform two additions, and compare the results. This can be done in constant time for every pair uv,xy. Since there are only $(n-1)m$ pairs of edges to be checked, Θ_1 (and thus Θ_1^*) $= \Theta^*$ can be computed within the required time.   □

Now we have all the prerequisites for the algorithm of Feder for the prime factorization of a connected graph with respect to the Cartesian product. It has complexity $O(mn)$ and is easy to implement. The algorithm of complexity $O(m)$ due to Imrich and Peterin [69], mentioned in Chapter 15, is much more complicated.

Theorem 18.6. [33] *For a connected graph G on n vertices and m edges, we can find the prime factorization with respect to the Cartesian product in $O(mn)$ time.*

Proof: By the discussion above, it remains to compute τ in $O(mn)$ time. We first consider all pairs e,f that have the unique common neighbor u. For every u, there are no more than $\frac{1}{2} \deg(u)^2$ such pairs; altogether, their number is at most

$$\frac{1}{2} \sum_{u \in G} \deg(u)^2 \leq \frac{n}{2} \sum_{u \in G} \deg(u) \leq nm.$$

Now, let $N(u)$ denote the neighborhood of u. For every vertex x of G, we determine $I = N(u) \cap N(x)$. For any two neighbors v, w in I, the pair vu, uw is not in τ. If $x \in N(u)$ and x has a neighbor $w \in N(u)$, then xu, uw is also not in τ. We remove these pairs. For every x, the complexity is $O(\deg(u))$. Since G has n vertices, the effort for u is $O(\deg(u)n)$. Summing over all $u \in G$, this yields $O(mn)$.   □

18.3   Recognizing Partial Cubes

Recall from Chapter 12 that isometric subgraphs of hypercubes are called partial cubes. In this section, we prove a characterization of partial cubes that is due to Winkler [118].

Theorem 18.7. [118] *A graph is a partial cube if and only if it is connected, bipartite, and $\Theta = \Theta^*$.*

Proof: Let G be an isometric subgraph of a hypercube Q_k. Clearly, G is connected; otherwise, it could not be isometric. G is bipartite because hypercubes are bipartite.

Let $\Theta(G)$ denote the relation Θ in G. For any two edges e, f in G, we infer by isometry that $e \, \Theta(G) \, f$ if and only if $e \, \Theta(Q_k) \, f$. Thus, $\Theta(G)$ is transitive because $\Theta(Q_k)$ is. (See Exercise 12.)

Now, let G be a connected, bipartite graph with transitive Θ. For any edge uv, let G_u (respectively G_v) denote the set of vertices of G that are closer to u (respectively v). Let xy be another edge with $uv \Theta xy$. Because G is bipartite, one of the vertices x, y has to be closer to u, and the other closer to v. (See Exercise 2.) Thus, all edges that are in the relation Θ to uv are between G_u and G_v, and no edge between G_u and G_v is in relation Θ to an edge with both endpoints in G_u or G_v.

Thus, the Θ^*-class of e, say Θ_e^*, consists of edges between G_u and G_v. By Exercise 4, the removal of a Θ^*-class disconnects a graph, and since G_u and G_v are connected, all edges between G_u and G_v must be in Θ_e^*. Thus $G - \Theta_e^*$ has two components, and $G/\Theta_e^* = K_2$.

Suppose the Θ^*-classes of G are $F_1, \ldots, F_k$. Since e was arbitrary, $G/F_i = K_2$ for all $i = 1, \ldots, k$. Hence the canonical metric representation

$$\alpha : G \to \square_{i=1}^{k} G/F_i$$

maps G into the hypercube Q_k. Now the assertion of the theorem follows, because α is an isometry by Theorem 14.4. $\square$

Corollary 18.8. *A connected, bipartite graph G is a partial cube if and only if the removal of any Θ^*-class decomposes G into two components.*

For every partial cube, the canonical isometric embedding is the unique irredundant isometric embedding into a hypercube.

By this corollary, it suffices to compute Θ^* and to show that the removal of any Θ^*-class decomposes G into two components in order to prove that G is a partial cube. By Proposition 18.5 one can determine Θ^* in $O(mn)$ time. Finding the number of components of every factor has complexity $O(m)$, since we have at most n factors, this part also has complexity $O(mn)$.

For partial cubes this really is $O(n^2 \log n)$, because subgraphs of hypercubes can have at most $O(n \log n)$ edges. This result is due to Graham [45]. For a proof see also Imrich and Klavžar [66].

Eppstein [31] reduced the complexity of recognizing partial cubes to $O(n^2)$. His main new ideas were to use bit-parallelism to speed up previous approaches to the embedding step, and to verify that the resulting embedding is isometric using an all-pairs shortest path algorithm he had previously developed.

We conclude that there are classes of partial cubes, in particular median graphs, that can be recognized faster; see Imrich and Klavžar [66].

18.4   Exercises

1. Show that incident edges are not in the relation Θ in bipartite graphs.

2. Let $e = uv, f = xy$ be edges of a bipartite graph G that are in the relation Θ. Let $d(u, x) \leq d(u, y)$. Show that $d(u, x) = d(v, y)$ and $d(u, y) = d(v, x) = d(u, x) + 1$.

3. Let G be a connected graph. Show that Θ^* has at most $|G| - 1$ equivalence classes.

4. Let Θ_e be the set of edges that are in relation Θ to e. Show that the graph $(V(G), E(G) - \Theta_e)$ is disconnected.

5. Let G be a connected graph and F be the union of one or more equivalence classes of Θ^*. Show that every connected component of the subgraph $H = (V(G), F)$ is convex.

6. Let H be a convex subgraph of G. Show that $\tau(H)$ is the restriction of $\tau(G)$ to H.

7. Let H be a G_i-fiber of the product $G = \square_{i=1}^{k} G_i$ of prime graphs. Show that the restriction of $\Theta^*(G)$ to H is $\Theta^*(H)$.

8. Let $\square_{i=1}^{k} G_i$ be the prime factorization of G and H a G_i-fiber of G. Show that $E(H)$ consists of only one $\sigma(H)$-class.

9. Let $G = \square_{i=1}^{k} G_i$ be a nontrivial product of connected graphs, and let v be a vertex of G. Show that $G - v$ is prime.

   (Hint: Removal of a vertex gives rise to new pairs of edges that are in the relation τ.)

10. Suppose G has a spanning subgraph isomorphic to $K_{2,3}$. Show that G has only one Θ^*-class.

11. [63] Show that the complement of two nontrivial graphs $G \square H$ is prime if one factor has more than four vertices.

12. Show that $\Theta(Q_k)$ is transitive.

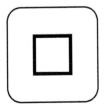

Hints and Solutions to Exercises

Part I—Cartesian Products

Chapter 1—The Cartesian Product

1. $|G \,\square\, H| = |G|\,|H|$.

2. $|E(G \,\square\, H)| = |G|\,|E(H)| + |H|\,|E(G)|$.

3. By the definition of the Cartesian product, $G \,\square\, K_1$ and $K_1 \,\square\, G$ are isomorphic to G; hence, K_1 is a unit.

   To see that the Cartesian product is commutative, we need to verify that the proposed mapping $G \,\square\, H \rightarrow H \,\square\, G$ defined with $(g, h) \mapsto (h, g)$ is a bijection that maps edges to edges and non-edges to nonedges. We show here that it maps edges to edges and leave the rest to the reader. Let $[(g, h)(g', h')]$ be an edge of $G \,\square\, H$. Suppose $g = g'$ and $[h, h'] \in E(H)$. Then (h, g) is adjacent to (h', g') in $H \,\square\, G$. We argue analogously for the case $[g, g'] \in E(G)$ and $h = h'$.

4. This is shown in a manner similar to Exercise 3: we need to show that the mapping $\phi : V((G_1 \,\square\, G_2) \,\square\, G_3) \rightarrow V(G_1 \,\square\, (G_2 \,\square\, G_3))$, defined by $\phi : ((g_1, g_2), g_3) \mapsto (g_1, (g_2, g_3))$, is an isomorphism. Clearly, ϕ is bijective, so we need to check that ϕ preserves adjacency and nonadjacency. Adjacent vertices $g = ((g_1, g_2), g_3)$ and $g' = ((g_1', g_2'), g_3')$ are different, and they differ in exactly one pair g_i, g_i', for which $[g_i, g_i'] \in E(G_i)$. But this condition also implies the adjacency of $\phi(g)$ and $\phi(g')$. If two or three of the pairs g_i, g_i', $1 \leq i \leq 3$, consist of two distinct elements, we have nonadjacency both of g and g' and of $\phi(g)$ and $\phi(g')$.

5. Select a vertex u of $K_3 \,\square\, K_3$ and its four nonneighbors. Then two of them that are not adjacent together with u form a K_3-fiber in

the complement of $K_3 \square K_3$, and similarly we find another K_3-fiber through u. The rest is now fixed. For instance, if we denote the vertices of $K_3 \square K_3$ with $[1,2,3;4,5,6;7,8,9]$, where this labeling is consistent with fibers (of the first factor), then a $K_3 \square K_3$ in the complement is induced on the vertices $[1,5,9;6,7,2;8,3,4]$.

6. First determine two triangles in the complement and the matching between them (that is, find the subgraph $K_2 \square K_3$ in the complement). Then complete the isomorphism.

7. The K_4-fibers are induced by the bipartition sets of $K_2 \square K_2 \square K_2$.

8. Hint: The intersection is the vertex (g,h).

9. Given a vertex $v = (g,h_1) \in G^{h_1}$, its neighbors are

$$\{(g',h_1) \mid gg' \in E(G)\} \cup \{(g,h) \mid hh_1 \in E(H)\}.$$

The first set is in G^{h_1}, the other one in ${}^gH$. Hence, v has the unique neighbor (g,h_2) in the fiber G^{h_2}.

Show now that the mapping $\varphi : G^{h_1} \to G^{h_2}$ defined by $\varphi : (g,h_1) \mapsto (g,h_2)$ is an isomorphism.

10. Suppose $G \square H - (x,y)$ is not connected. Let (g,h) and (g',h') be vertices from different connected components of $G \square H - (x,y)$. If $g \neq g'$ and $h \neq h'$, then, using the notation of Lemma 1.1, we see that $(R \times \{h\}) \cup (\{g'\} \times S)$ and $(\{g\} \times S) \cup (R \times \{h'\})$ induce internally disjoint $(g,h),(g',h')$-paths, which is not possible. Thus, we assume without loss of generality that $g = g'$ and $h \neq h'$. Let g'' be an arbitrary neighbor of g in G. But then $\{g\} \times S$ and $(g,h) \cup (\{g''\} \times S) \cup (g,h')$ induce internally disjoint $(g,h),(g',h')$-paths, another contradiction.

11. Let V_1, V_2 be a bipartition of the bipartite graph G and let W_1, W_2 be a bipartition of the bipartite graph H. Then

$$(V_1 \times W_1) \cup (V_2 \times W_2), (V_1 \times W_2) \cup (V_2 \times W_1)$$

is a bipartition of $G \square H$.

12. $L(C_n)$ is connected and has n vertices, each of degree two.

13. A vertex of degree $d \geq 3$ yields a complete subgraph K_d in $L(G)$, so $L(G)$ is not a tree. However, $L(P_n) = P_{n-1}$.

14. A path between two vertices e and f of the line graph is naturally obtained from a path between an end vertex of e and an end vertex of f in the original graph.

15. Let e be a vertex of a line graph adjacent to f_1, f_2, and f_3. Then in the original graph, at least two of the edges f_1, f_2, and f_3 must be incident to the same end vertex of e, say f_1 and f_2. It follows that f_1 and f_2 are adjacent in the line graph.

16. It suffices to show that any two vertices that are the endpoints of a walk of length k are adjacent. By assumption, this is true for $k = 2$. Assume that it is true for $k - 1 \geq 2$ and let $g_0 g_1 \ldots g_{k-1} g_k$ be a walk. Then $g_0 g_{k-1} \in E(G)$ by the induction assumption. Hence, $g_0 g_{k-1} g_k$ is a walk, and thus g_0 and g_k are adjacent.

Chapter 2—Hamming Graphs and Hanoi Graphs

1. Q_3 is isomorphic to the graph obtained from $K_{4,4}$ by removing a perfect matching.

2. Let $G = K_{n_1} \square K_{n_2} \square \cdots \square K_{n_k}$, $n_i \geq 2$, be a Hamming graph. If for some i, $n_i \geq 3$, then the K_{n_i}-fibers demonstrate that G is not bipartite. So in order for G to be bipartite, $n_i = 2$ must hold for every i, and consequently $G = Q_k$.

3. The prism over Q_n is Q_{n+1}.

4. **First solution:** Let u be an arbitrary vertex of Q_n, say $u = 00 \ldots 0$. Then u has n neighbors. Each pair of neighbors of u, say $100 \ldots 0$ and $010 \ldots 0$, gives precisely one 4-cycle containing u. In our case, it is induced by u, $100 \ldots 0$, $010 \ldots 0$, and $110 \ldots 0$. Since there are $\binom{n}{2}$ pairs of neighbors of u, there are precisely that many 4-cycles containing u. This argument holds for any vertex of Q_n. There are 2^n vertices and every square is counted four times, so together there are

$$\frac{2^n \binom{n}{2}}{4} = 2^{n-2} \binom{n}{2}$$

4-cycles.

Second solution: Select two coordinates i, j in the coordinate vector of Q_n and fix all the other coordinates. There are four possibilities to place zeros and ones into the selected two coordinate positions. The corresponding four vertices induce a 4-cycle. Moreover, any 4-cycle is of such a form. Since two coordinates can be selected in $\binom{n}{2}$ ways and since there are 2^{n-2} possibilities to fix the remaining coordinates, the number of 4-cycles is $2^{n-2} \binom{n}{2}$.

5. Two adjacent vertices differ in exactly one coordinate. Hence, two vertices u, v that differ in k coordinates cannot be the endpoints of a path having fewer than k edges. In other words, their distance is at least k. Complete the proof by showing that u and v can be connected by a path of length k.

6. Use the same argument as in Exercise 5.

7. Let $X = \{x_1, x_2, \ldots, x_n\}$ and let $A \subseteq X$. Then assign to A the binary n-tuple $(a_1, a_2, \ldots, a_n)$, where $a_i = 1$ if $x_i \in A$, and $a_i = 0$ otherwise. It is now straightforward to verify that these n-tuples define the vertex set of Q_n.

8. The Hanoi graph H_4^2 is drawn in Figure 1.

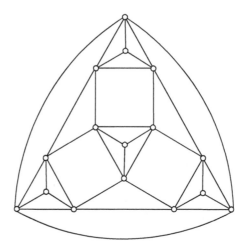

Figure 1. The Hanoi graph H_4^2.

The reader is invited to complete the figure by labeling its vertices.

9. The statement is clearly true for $n = 1$. Suppose now that it holds for $n \geq 1$ discs, and consider the puzzle with $n + 1$ discs. It can be solved as follows. First transfer the topmost n discs from peg 1 to peg 2, then move the largest disc from peg 1 to peg 3, and complete the solution by moving the n discs from peg 2 to peg 3. By the induction assumption this solution requires $(2^n - 1) + 1 + (2^n - 1) = 2^{n+1} - 1$ moves.

Part II—Classic Topics

Chapter 3—Hamiltonian Graphs

1. Every cycle in a bipartite graph has even order.

2. If neither n nor m is even, then $P_n \square P_m$ is a bipartite graph of odd order and is not hamiltonian by Exercise 1. For the converse, assume that n is even, let $V(P_n) = \{1, 2, \ldots, n\}$, and let $V(P_m) = \{1, 2, \ldots, m\}$. Let A_i, $1 \le i \le n$, denote the path of order $m - 1$ induced by the vertices $\{(i, j) \mid 2 \le j \le m\}$, and let F be the subgraph of $P_n \square P_m$ that is the union of the subgraphs $A_1, A_2, \ldots, A_n$. A hamiltonian cycle is obtained by adding to F the fiber P_n^1 along with the sets of edges

   - $\{[(j, 2)(j + 1, 2)] \mid 2 \le j \le n - 2, j \text{ even}\}$,
   - $\{[(1, 1)(1, 2)], [(n, 1)(n, 2)]\}$, and
   - $\{[(j, m)(j + 1, m)] \mid 1 \le j \le n - 1, j \text{ odd}\}$.

3. Let A_1, A_2 be the bipartition of G and let B_1, B_2 be the bipartition of H such that $n_1 = |A_1| \ge |A_2| = n_2$ and $m_1 = |B_1| \ge |B_2| = m_2$. The graph $G \square H$ is bipartite with bipartition $(A_1 \times B_1) \cup (A_2 \times B_2)$ and $(A_1 \times B_2) \cup (A_2 \times B_1)$. Computing the cardinalities of these sets, we get

$$|(A_1 \times B_1) \cup (A_2 \times B_2)| - |(A_1 \times B_2) \cup (A_2 \times B_1)| = (n_1 - n_2)(m_1 - m_2),$$

   which is nonzero if neither G nor H is color-balanced. Since every cycle in a bipartite graph contains the same number of vertices from each part of the bipartition, it follows that $G \square H$ is not hamiltonian if neither is color-balanced.

4. Hint for $n = 4$: Let $\{a\} \cup \{u, v, w, x\}$ be the bipartition of $V(K_{1,4})$, and let $1, 2, 3, 4, 5, 6, 7, 8, 1$ be the cycle of order eight. A spanning cycle for $K_{1,4} \square C_8$ begins $(u, 1), (u, 2), \ldots, (u, 8), (a, 8), (a, 7), (v, 7), (v, 8), (v, 1)$.

5. Suppose $g \in G$ is adjacent to g_1 and g_2 of degree one, and $h \in H$ is adjacent to h_1 and h_2 of degree one. In $G \square H$ the four vertices (g_i, h_j) for $i, j \in \{1, 2\}$ each have degree two. The only cycle of $G \square H$ that contains all four of these vertices has order eight, and $G \square H$ has order at least nine.

6. Assume that for every tree T of order at most n, there is an edge coloring $f : E(T) \to \{1, 2, \ldots, \Delta(T)\}$. Assume that S is a tree

of order $n + 1$ and that x is a vertex of S such that $\deg_S(x) = \Delta(S) = k$. For an edge $e = xy$, $S - e$ has two components S_1 and S_2 that are trees such that $x \in S_1$ and $y \in S_2$. It is clear that each of S_1 and S_2 has maximum degree at most k. By the induction assumption, there are edge colorings c_1 and c_2 of S_1 and S_2, respectively, that each use a maximum color no larger than k. Since $\deg_{S_1}(x) < k$ and $\deg_{S_2}(y) < k$, we may assume—by permuting colors, if necessary—that no edge incident with x is assigned color k by c_1. A similar property can be assumed about y and c_2. By defining c as the union of c_1 and c_2 and setting $c(e) = k$, we get the desired edge coloring of S. The result follows by induction.

A simpler solution that actually produces the edge coloring is as follows. Let $\Delta(T) = k$. Root T at an arbitrary vertex and proceed down the tree using the greedy algorithm to color the edges. When a vertex v is encountered, exactly one edge incident with that vertex has been colored, and thus there are $k - 1$ other colors available to color the remaining $\deg(v) - 1 \leq k - 1$ edges incident with v.

7. Assume that G has order n and H has order m. Since G and H are traceable, the complete grid graph $P_n \,\square\, P_m$ is a spanning subgraph of $G \,\square\, H$. By Exercise 2, this spanning subgraph has a hamiltonian cycle C as long as at least one of n or m is even. Hence, in this case, C is a hamiltonian cycle of $G \,\square\, H$ as well. Assume then that both of n and m are odd, but at least one of the two graphs, say H, has an odd cycle F. We assume that G is a path of order n and that H is a path of order m with a single additional edge to create the odd cycle F. The two cases in Figure 2 show how to construct a spanning cycle of $G \,\square\, H$, depending on whether the removal of the edges of F from H leaves two paths of odd order or two paths of even order in addition to some isolated vertices.

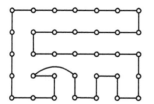 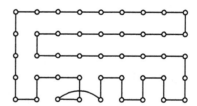

Figure 2. Examples of two cases. Two paths of odd order (left) and two paths of even order (right) after removal of edges of F.

8. Since G and H are hamiltonian, they are traceable. If at least one of $n = |G|$ or $m = |H|$ is even, it follows from Exercise 7 that $G \square H$ is hamiltonian. For the case when both n and m are odd, we proceed in a manner similar to that followed in Exercise 2. Assume that the spanning cycle of G is $g_1, g_2, \ldots, g_n, g_1$ and that of H is $h_1, h_2, \ldots, h_m, h_1$. Let A_i, $1 \leq i \leq n$, denote the path of order $m - 1$ induced by the vertices $\{(g_i, h_j) \mid 2 \leq j \leq m\}$, and let F be the subgraph of $G \square H$ that is the union the subgraphs $A_1, A_2, \ldots, A_n$. A hamiltonian cycle is obtained by adding to F the fiber G^{h_1} along with the edges:

- $[(g_j, h_2)(g_{j+1}, h_2)]$ for $2 \leq j \leq n - 1, j$ even,
- $[(g_1, h_1)(g_1, h_2)], [(g_n, h_1)(g_n, h_m)]$, and
- $[(g_j, h_m)(g_{j+1}, h_m)]$ for $1 \leq j \leq n - 2, j$ odd.

9. Note first that if x is a leaf of T, then by its very definition, the successor function s produces a directed spanning path in the fiber $^x H$. For example, if the edge xy incident with x is colored k and $x \in V_1$, then

$$(x, h_{k+1}), (x, h_{k+2}), \ldots, (x, h_n), (x, h_1), \ldots, (x, h_k)$$

is the directed path determined by s.

Let $T' = T - x$. Assume without loss of generality that x and y are as above and let s' be the successor function as defined in the proof of Theorem 3.4 based on the coloring of T' induced by the original edge coloring of T. Then, by induction, s' produces a spanning cycle C' of $T' \square H$, and $s'(y, h_{k+1}) = (y, h_k)$. The function s agrees with s' on $T' \square H$, except that $s(y, h_{k+1}) = (x, h_{k+1})$. In addition, $s(x, h_i) = (x, h_{i+1})$ for $i \neq k$ and $s(x, h_k) = (y, h_k)$. That is, the directed path

$$(x, h_{k+1}), (x, h_{k+2}), \ldots, (x, h_n), (x, h_1), \ldots, (x, h_k)$$

has been inserted between (y, h_{k+1}) and (y, h_k) in C'. The result is a spanning cycle of $T \square H$. The general result now follows by induction on $|T|$.

10. Let g_1, g_2, g_3 be vertices of degree one in G with common neighbor g and let h be a vertex of degree one in H. In $G \square H$, the vertices $(g_1, h), (g_2, h)$, and (g_3, h) are of degree two, and all three are adjacent to the vertex (g, h). Any cycle of $G \square H$ can have at most two edges incident with (g, h) and thus can contain at most two of $(g_1, h), (g_2, h)$, and (g_3, h).

11. Let $\{1,2\} \cup \{u_1, v_1, \ldots, u_m, v_m\}$ be the bipartition of $K_{2,2m}$. The set of edges $\{1u_1, 1u_2, \ldots, 1u_m, 2u_m, 2v_1, \ldots, 2v_m\}$ forms a spanning tree of $K_{2,2m}$, and this shows that $\nabla(K_{2,2m}) \leq m + 1$. If T is any spanning tree, then $|N_T(1) \cap N_T(2)| = 1$ and $|N_T(1) \cup N_T(2)| = 2m$. This implies that either $|N_T(1)|$ or $|N_T(2)|$ is at least $m + 1$, and hence $\nabla(K_{2,2m}) \geq m + 1$.

(Hint for $m = 3$: Let $\{a, b\} \cup \{u_1, u_2, \ldots, u_6\}$ be the bipartition of $K_{2,6}$, and let $1, 2, 3, 1$ be the 3-cycle. A spanning cycle of $K_{2,6} \,\square\, C_3$ begins

$$(u_1, 1), (u_1, 2), (u_1, 3), (a, 3), (u_2, 3), (u_2, 1), \ldots).$$

12. Let the bipartition of $K_{2,4}$ be $\{u_1, u_2\}$, $\{v_1, v_2, v_3, v_4\}$ and let $V(K_2) = \{1, 2\}$. The list $(u_1, 1)$, $(v_1, 1)$, $(v_1, 2)$, $(u_1, 2)$, $(v_2, 2)$, $(v_2, 1)$, $(u_2, 1)$, $(v_3, 1)$, $(v_3, 2)$, $(u_2, 2)$, $(v_4, 2)$, $(v_4, 1)$, $(u_1, 1)$ is a spanning cycle of the prism $K_{2,4} \,\square\, K_2$.

13. Let $\{a, b\}$ and $\{u, v, w, x, y\}$ be a bipartition of $K_{2,5}$, and let $V(K_2) = \{1, 2\}$. If the four vertices $(a, 1), (a, 2), (b, 1), (b, 2)$ are removed from $K_{2,5} \,\square\, K_2$, the resulting graph has five components. By Lemma 3.2, it follows that $K_{2,5} \,\square\, K_2$ is not hamiltonian.

Chapter 4—Planarity and Crossing Number

1. Let $G_1, \ldots, G_r$ and $H_1, \ldots, H_s$ be the connected components of G and H, respectively. Then the connected components of $G \,\square\, H$ are $G_i \,\square\, H_j$, $1 \leq i \leq r$, $1 \leq j \leq s$.

2. There are only six vertices in $K_{1,3} \,\square\, P_3$ that have degree at least three.

3. Consider $G \,\square\, H$. Suppose first that both G and H have at least three vertices. Then $P_3 \,\square\, P_3$ is a subgraph of $G \,\square\, H$. Since $P_3 \,\square\, P_3$ is not outerplanar (find a subdivision of $K_{2,3}$!), neither is $G \,\square\, H$. Suppose next $G = K_2$ and H is not a tree. Then $K_2 \,\square\, C_n$ is a subgraph of $G \,\square\, H$ that is not outerplanar. (Verify it!) And if H is a tree that is not a path, then $K_2 \,\square\, K_{1,3}$ is a nonouterplanar subgraph (verify it!) of $G \,\square\, H$.

The natural drawing of $P_n \,\square\, K_2$ shows its outerplanarity.

4. Let $V(K_2) = \{a, b\}$. Consider the (copy of the) subdivision of H in the fiber G^a; denote it with A. Then there exist internally disjoint paths from the vertex (u, b) to all the vertices in A that correspond to $N[u]$.

5. Since K_5 and $K_{3,3}$ are the critical nonplanar graphs, it is no surprise that $\mathrm{cr}(K_5) = \mathrm{cr}(K_{3,3}) = 1$. Draw corresponding figures!

6. Draw five concentric circles, and perpendicular to them draw the three $K_{2,3}$-fibers. Then each fiber will give 4 crossings. (Note that by Theorem 4.9, this drawing is not optimal. For such a drawing we refer to Klešč [82].)

7. The formula suggests that we can draw a $K_{1,m}$-fiber vertically in such a way that the center of the star is the middle vertex. Then put all the $K_{1,m}$-fibers in parallel and connect them with the P_n-fibers.

8. Consider Figure 3, where the first two $(K_2 \square K_3)$-fibers are drawn. (Check that the fibers are indeed $K_2 \square K_3$!)

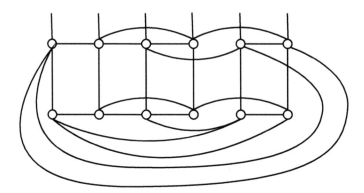

Figure 3. First two $(K_2 \square K_3)$-fibers of a drawing of $P_n \square (K_2 \square K_3)$.

Continue drawing $(K_2 \square K_3)$-fibers as indicated, and finally draw the last fiber symmetric to the first one. Then each of the first and the last fibers gives 2 crossings, whereas any of the middle $n - 2$ gives 4 crossings. Hence, there are $4(n - 1)$ crossings altogether.

Chapter 5—Connectivity

1. Let G_1 and G_2 be two connected components of $G - S$. (There may be more connected components, but we need to consider only two of them.) Then $G_1 \square H$ and $G_2 \square H$ contain at least two connected components of $(G \square H) - (S \times V(H))$. (If H is connected, then there are exactly two such components.) Hence $S \times V(H)$ is a separating set of $G \square H$.

2. Let G_n be the graph that is obtained from two copies of K_n by identifying a vertex of one copy with a vertex of the other copy. Then verify that any pair of graphs G_n, G_m, where $n \geq 3$ and $m > n + 1$, does the job.

3. There are several ways to show that $\kappa(P_m \square P_n) = 2$. It can be shown directly by finding, for each pair of vertices, two internally disjoint paths connecting them. As another option, we can also apply Theorem 5.1. The reader familiar with the notion of ear decompositions can also use this technique.

4. For the first part, use Theorem 5.1 and note that $\delta(G) + 1 \leq |G|$. For the second part, use the graphs G_n from the solution of Exercise 2 and complete graphs.

5. Use induction on n and the fact that $Q_n = Q_{n-1} \square K_2$. The key observation is that a separating set of Q_n must separate both Q_{n-1}-fibers.

   Another approach would be to find n internally vertex-disjoint paths between each pair of vertices of Q_n.

6. Analogous to the case of the vertex connectivity, we can easily infer that $\kappa'(P_m \square P_n) = 2$. (See also the next exercise!)

7. Observe that internally vertex-disjoint x, y-paths are also internally edge-disjoint x, y-paths. Then apply Menger's theorem for vertices and Menger's theorem for edges. (See West [115] for an exposition of Menger's Theorem.)

8. Consider a minimum separating set S of G. Since in general $\kappa'(G) \geq \kappa(G)$ (Exercise 7), we need to find a disconnecting set of edges of size $|S|$. Let G_1 and G_2 be two connected components of $G - S$. Since S is minimum, every vertex $u \in S$ has a neighbor in G_1 as well as a neighbor in G_2. But since G is 3-regular, either u has exactly one neighbor in G_1 or exactly one neighbor in G_2. Now it is not difficult to determine the required disconnecting set of edges (there are a couple of cases to be considered).

9. Suppose it is not. Then find a path in $G \square H$ whose projection to G contradicts the assumption that S is a disconnecting set of edges in G.

10. For $k = 2$, this is Theorem 5.5; for $k \geq 3$, proceed by induction.

Chapter 6—Subgraphs

1. Let X be a nontrivial Cartesian subgraph of $G \square H$. Let $V(G) = \{g_1, \ldots, g_n\}$ and $V(H) = \{h_1, \ldots, h_m\}$. Draw $G \square H$ such that the vertex (g_i, h_j) is represented with the point (i, j), $1 \le i \le n$, $1 \le j \le m$. The corresponding subdrawing of X is then a nontrivial plotting.

   Assume X has a nontrivial plotting. Then every edge of X projects on one of the coordinate axes. The two projections give the desired factor graphs.

2. Let v be a cut vertex of G and draw it as the point $(0, 0)$. Let H be a connected component of $G - v$. Then draw the vertices of H as different points with the first coordinate 0, and the vertices of $G - H$ as different points with the second coordinate 0. Apply Exercise 1.

3. Draw the vertices of A in one horizontal level and the vertices of B in another horizontal level such that the matching edges between A and B are all vertical. Then every other edge of G will be horizontal.

4. We will verify condition (ii) of Theorem 6.1. First we consider the graphs from Exercise 2. Let v be a cut vertex of G and let H be a connected component of $G - v$. Then label the edges of H and the edges between v and H with 1 and all the other edges with 2. Then every cycle of G is labeled with one color only and the condition is trivially fulfilled.

   Let G be a graph on at least three vertices with a partition $\{A, B\}$ of $V(G)$, as in Exercise 3. Then label the matching edges with 1 and the remaining edges with 2.

   Alternatively, the reader is invited to verify condition (iii) of Theorem 6.1.

5. Let G be a bipartite graph with radius 2, and suppose that G contains no subgraph isomorphic to $K_{2,3}$. Let v be a vertex of G such that all vertices of G are at distance at most two from it, and let $vw \in E(G)$. Label w and all its neighbors but v with 1 and all the other vertices with 2. Hence v is labeled 2, and since G is bipartite, any neighbor of v, except w, is also labeled 2.

   We claim that the above labeling fulfills condition (iii) of Theorem 6.1. Suppose not, so that there exists a properly colored path xyz of G for which $c(x) = c(z) \ne c(y)$. Assume first $c(x) = 1$.

If $x = w$, then $y = v$, and since v has only one neighbor labeled 1, we cannot find z on P. It follows that $x \neq w$. Since G is bipartite, y is adjacent to v and z is adjacent to w. So w, y, v, x, z induce a $K_{2,3}$. Assume next that $c(x) = 2$. Then x is not adjacent to w, and using the bipartiteness we again infer that x is adjacent to v. Then y is adjacent to w and z is adjacent to v. Again, v, y, w, x, z induce a $K_{2,3}$.

6. Let x and y be different vertices of X. We first show injectivity of f, that is, $f(x) = (\ell_V(x), g(x)) \neq (\ell_V(y), g(y)) = f(y)$. If $\ell_V(x) \neq \ell_V(y)$ we are done. Hence, suppose $\ell_V(x) = \ell_V(y)$. Then no x, y-path is properly colored by assumption (iii) from the theorem. Therefore, x and y are in different connected components of Z, which in turn implies that $g(x) \neq g(y)$.

   Let $xy \in E(X)$. We need to show that $f(x)f(y) \in E(K_k \square K_r)$. Now, if $\ell_V(x) = \ell_V(y)$, then we have seen above that $g(x) \neq g(y)$; hence, the desired conclusion. If $\ell_V(x) \neq \ell_V(y)$, then clearly $g(x) = g(y)$, and we are done.

7. Systematically consider all the subgraphs on at least three vertices. For each of them, it is easily inferred to be a nontrivial Cartesian subgraph.

8. Let $K_n \subseteq G \times H$ and, as a subgraph of $G \times H$, let K_n be induced on the vertices (g_i, h_i), $1 \le i \le n$. Consider vertices (g_j, h_j) and (g_k, h_k), where $j \neq k$. Since (g_j, h_j) and (g_k, h_k) are adjacent in $G \times H$, by the definition of the direct product, we find that $g_j \neq g_k$ and $h_j \neq h_k$. Consequently, the vertices $g_1, \ldots, g_n$ induce a K_n in G and $h_1, \ldots, h_n$ induce a K_n in H.

Part III—Graphical Invariants

Chapter 7—Independence

1. Let $V_1, V_2, \ldots, V_n$ be a partition of $V(G)$ into independent sets. The set $\cup_{j=1}^n (\{j\} \times V_j)$ is independent in $K_n \square G$; we conclude that $\alpha(K_n \square G) \ge |G|$.

   Now the pigeonhole principle, together with the observation that the K_n-fibers are complete, implies $\alpha(K_n \square G) \le |G|$.

2. Simply use the definitions of dominating set and maximal independent set.

3. If A is a maximum independent set, then every edge of G is incident with at least one vertex of $V - A$, which implies that $\beta(G) \leq |V - A| = |G| - \alpha(G)$. On the other hand, if B is a minimum vertex cover, then no edge is induced by $V - B$. That is, $V - B$ is independent. Now, $\alpha(G) \geq |V - B| = |G| - \beta(G)$.

4. It is clear from the construction that at least two vertices from each of the original 5-cycles must be included in any maximal independent set of H_n. However, an independent set can intersect a 5-cycle at most twice. It now follows that $i(H_n) = 2n = \alpha(H_n)$.

5. Let A be a subset of $V(G)$ of cardinality $\alpha_n(G)$, and let $A_1, A_2, \ldots, A_n$ be a weak partition (that is, a partition in which some parts are allowed to be empty) of A into independent sets. The set $\cup_{j=1}^n (A_j \times \{j\})$ is independent in $G \square K_n$. Any maximum independent set I in $G \square K_n$ can intersect each K_n-fiber in at most one vertex. Thus, if $I_j = P_G(I \cap G^j)$, then

$$\alpha_n(G) \geq |I_1 \cup I_2 \cup \cdots \cup I_n| = |I| = \alpha(G \square K_n).$$

6. For (1), use the pigeonhole principle and the fact that every G-fiber is isomorphic to G and every H-fiber is isomorphic to H. For any edge $h_1 h_2$ of H, it follows from Exercise 5 that the subgraph induced by $V(G) \times \{h_1, h_2\}$ contains at most $\alpha_2(G)$ vertices from any independent set of $G \square H$. Now consider G-fibers corresponding to vertices that remain when the vertices saturated by a maximum matching in H are removed.

7. Select G_1 to play the role of H in Corollary 7.5(ii).

8. Follow the example preceding Theorem 7.6 to construct an independent set of cardinality $\alpha(G)|H|$ if there is a homomorphism $H \to \mathrm{Ind}(G)$. For the converse, consider how a maximum independent set of $G \square H$ of cardinality $\alpha(G)|H|$ intersects two G-fibers corresponding to an edge from H. Use the projection maps to define a homomorphism $H \to \mathrm{Ind}(G)$.

9. What is $\mathrm{Ind}(P_6)$? Consider Theorem 7.6 in this particular context.

10. Let M be any maximum independent set of $G \square P_{2n+1}$ and denote $M \cap G^{v_j}$ by M_j. Select k such that $|M_k| + |M_{k+1}| \geq |M_j| + |M_{j+1}|$ for every $1 \leq j \leq 2n$. Let A be $p_G M_k$ or $p_G M_{k+1}$, based on which has a larger cardinality. How should B be chosen? Now show I is independent and that $|I| \geq |M|$.

11. Use Proposition 7.3 to compute $\alpha(P_{2r} \square P_m) = rm =$ $\alpha(C_{2r} \square P_m)$, and $\alpha(P_{2r} \square C_m) = 2r\lfloor \frac{m}{2} \rfloor = \alpha(C_{2r} \square C_m)$. Show that $\alpha(P_{2r+1} \square P_{2s+1}) = 2rs + r + s + 1$ by combining Proposition 7.3 and Corollary 7.5. If $s \geq r$, $\alpha(C_{2r+1} \square C_{2s+1}) = r(2s + 1)$. For any r and s, $\alpha(P_{2r+1} \square C_{2s+1}) = s(2r + 1)$.

Chapter 8—Graph Colorings

1. If no such edge xy exists, then there is a homomorphism $G \rightarrow K_{n-1}$.

2. Use a $\chi(G)$-coloring of G and a $\chi(H)$-coloring of H to show $\chi(G \vee H) \leq \chi(G) + \chi(H)$. Then prove that if $c : G \vee H \rightarrow K_n$ is a homomorphism, $c^{-1}(i) \subseteq V(G)$ or $c^{-1}(i) \subseteq V(H)$ for every i to verify the reverse inequality.

3. If, for example, $\chi(G \square H) = \chi(G)$ by Theorem 8.1, then every G-fiber has the same chromatic number as $G \square H$.

4. Use the definitions of $\chi(G)$, $\psi(G)$, and greedy coloring.

5. Order the first four vertices from one end of P_n in such a way that forces a greedy coloring to use three colors, and then order the remaining $n - 4$ vertices so that the greedy coloring does not use more than three.

6. Let $V_1, V_2, \ldots, V_{\chi(G)}$ be a coloring of G. Design an appropriate ordering based on this coloring.

7. Consider a star with $\binom{n}{2}$ leaves.

8. Assume that $V_1, V_2, \ldots, V_k$ is a complete coloring of G and let $V(P_2) = \{1, 2\}$. Define $c : G \square P_2 \rightarrow \{1, 2, \ldots, k + 1\}$ such that $c^{-1}(1) = V_1 \times \{1\}$, $c^{-1}(k + 1) = V_1 \times \{2\}$, $c^{-1}(k) = (V_k \times \{1\}) \cup (V_2 \times \{2\})$, and $c^{-1}(j) = (V_j \times \{1\}) \cup (V_{j+1} \times \{2\})$ for $2 \leq j \leq k - 1$.

9. Use the ordering $u_1, v_1, u_2, v_2, \ldots, u_n, v_n$, where $\{u_i, v_i\}_{i=1}^{n}$ is the perfect matching that is removed.

10. Choose an ordering of $V(G)$ whose last vertex has degree less than $\Delta(G)$ in such a way that the greedy coloring with respect to this ordering does not require more than $\Delta(G)$ colors.

11. Each part in a $\mathcal{D}_1$-partition of K_n can have at most two vertices.

12. By Theorem 8.1, $\chi(G) = 2$, and $\chi_{\mathcal{D}_1}(G)$ cannot be 1.

13. By Theorem 8.6, $\chi_{\mathcal{I}_k}(G \square H) \geq \max\{\chi_{\mathcal{I}_k}(G), \chi_{\mathcal{I}_k}(H)\}$ since $\mathcal{I}_k$ is a hereditary property. Now prove that if $G, H \in \mathcal{I}_k$, then neither $G \square H$ nor the disjoint union of G and H has a complete subgraph of order $k + 2$.

14. Let C_r denote a circle of circumference $r > 0$ with center at the origin, and let P denote the point where C_r intersects the horizontal axis. We identify C_r with the interval $[0, r)$ of real numbers in such a way that P is identified with 0. For any other point Q of C_r we find the arc length from P to Q measured clockwise on C_r. If this length is s, then Q is identified with s. Using this identification, an r-circular coloring of a graph G can be thought of as a map c from $V(G)$ to the interval $[0, r)$ that assigns to a vertex g of G the real number identified with the beginning of the arc as it is laid out on the circle C_r in a clockwise direction.

    Now, suppose that $c : G \to [0, k/d)$ is a (k/d)-circular coloring of G. We must convince ourselves that $c' : G \to Z_k$ defined by $c'(g) = \lfloor c(g)d \rfloor$ is a (k, d)-coloring of G. Conversely, for a (k, d)-coloring f of G, the map defined by $f'(g) = f(g)/d$ is a (k/d)-circular coloring of G. Check this!

    How does this verify Equation (8.2)?

15. Check the condition separately on edges, depending on whether they belong to a G-fiber or an H-fiber.

Chapter 9—Additional Types of Colorings

1. Find an ordering $v_{\pi(1)}, \ldots, v_{\pi(n)}$ of $V(T)$ such that for each $1 < i \leq n$, $v_{\pi(i)}$ is a leaf of the subtree of T that is induced by $\{v_{\pi(1)}, \ldots, v_{\pi(i)}\}$.

    If G is regular, then every permutation π yields the same value for $\max_i\{d_\pi(v_{\pi(i)})\}$.

2. Assume that the vertices of G have been labeled $g_1, \ldots, g_n$ such that $\mathrm{col}(G) = 1 + \max_i\{d(g_i)\}$, where $d(g_i)$ denotes the number of neighbors of g_i in $\{g_1, \ldots, g_{i-1}\}$. Similarly, assume the vertices of H have been labeled $h_1, \ldots, h_m$ such that $\mathrm{col}(H) = 1 + \max_i\{d(h_i)\}$. Use the lexicographic ordering of $V(G \square H)$,

$$(g_1, h_1), \ldots, (g_1, h_m), (g_2, h_1), \ldots,$$
$$(g_2, h_m), \ldots, (g_n, h_1), \ldots, (g_n, h_m).$$

The vertex (g_i, h_j) has at most $d(g_i) + d(h_j) \leq \mathrm{col}(G) - 1 + \mathrm{col}(H) - 1$ neighbors preceding it in this list.

3. If $\{a, b\}, \{u, v, w, x\}$ is the bipartition of $K_{2,4}$, use list assignments $L(a) = \{1, 2\}$, $L(b) = \{3, 4\}$, $L(u) = \{1, 3\}$, $L(v) = \{1, 4\}$, $L(w) = \{2, 3\}$, and $L(x) = \{2, 4\}$.

4. Assume $\chi_\ell(G) = k$. If L is a list assignment of H of order k, then L can be extended (in an arbitrary manner) to a list assignment L' of G of order k. But G has a list coloring $c : G \to K_n$ corresponding to L'. The restriction of this homomorphism to the subgraph H will be a list coloring of H corresponding to L. Hence, $\chi_\ell(H) \leq k$.

5. By Exercise 1, $\mathrm{col}(P_2) = 2 = \mathrm{col}(P_3)$, and thus, by Proposition 9.2, we get $\chi_\ell(P_2) = 2 = \chi_\ell(P_3)$. This gives an upper bound of 3 by Theorem 9.3. The reader should find a specific list assignment of order two for $P_2 \,\square\, P_3$ that has no corresponding list coloring.

6. Let $V_1, V_2, \ldots, V_{\chi(G)}$ be a coloring of G. Label the vertices in V_1 in any order using labels $0, 1, \ldots, |V_1| - 1$. Use labels $|V_1| + 1, |V_1| + 2, \ldots, |V_1| + |V_2|$ to label the vertices of V_2 in any order. Continue this labeling strategy. Show this is an $L(2, 1)$-labeling and that the maximum label used is $|G| + \chi(G) - 2$.

7. $\lambda_{2,1}(P_6) = 4 = \lambda_{2,1}(C_6)$,
   $\lambda_{2,1}(K_8) = 14$,
   $\lambda_{2,1}(K_{3,3}) = 6 = \lambda_{2,1}(P_3 \,\square\, P_3)$.

8. For every pair $1 \leq i < j \leq n$, $|L(i) - L(j)| \geq 2$ for any $L(2, 1)$-labeling of K_n. This implies that $\lambda_{2,1}(K_n) \geq 2n - 2$. Conversely, $L(k) = 2k - 2$ defines an $L(2, 1)$-labeling of K_n.

9. See the paragraph just before Proposition 9.5.

10. The fact that $K_n \,\square\, K_2$ has diameter 2 and order $2n$ implies that

$$\lambda_{2,1}(K_n \,\square\, K_2) \geq 2n - 1.$$

For the reverse inequality, verify that the labeling given by $f(k, 1) = 2k - 2$ and $f(k, 2) = 2n - 2k + 1$ is an $L(2, 1)$-labeling.

11. Let u, v, w be vertices of degree Δ with u adjacent to both v and w, and suppose that L is an $L(2, 1)$-labeling of G with maximum label $\Delta + 1$. If $L(u) \neq 0$, then $L(x) \geq \Delta + 2$ for some neighbor x of u. Now it follows that one of $L(v)$ or $L(w)$ is in $\{2, 3, \ldots, \Delta\}$, which leads directly to a contradiction.

12. It is easily verified directly that the labeling of $C_4 \,\square\, C_4$ given in Figure 9.1 is an $L(2, 1)$-labeling. Let u be an arbitrary vertex of

$C_4 \square C_4$. Then the graph induced by the vertices that are at a distance more than two from u is isomorphic to $K_{1,4}$. (The vertex that is diametrical to u is the center of the star $K_{1,4}$.) It follows that at most two vertices can receive the same label. Since $C_4 \square C_4$ has 16 vertices, we need at least 8 different labels.

13. For the upper bound, form a breadth-first spanning tree rooted at any vertex, label that vertex 0, and proceed in a greedy fashion.

14. Suppose $v_1, v_2, \ldots, v_n$ is a spanning path of $\overline{G}$. In particular, this means that $v_i v_{i+1}$ is not an edge in G. Label v_1 by 0. How can this be extended to an $L(2,1)$-labeling of G?

15. This is an immediate application of Proposition 9.7. A more instructive solution can be constructed by labeling, in a consecutive fashion, the members of an entire partite set. Then proceed to the next partite set, and so forth.

16. For an r-regular graph, a 1-factorization corresponds, by its very definition, to an edge coloring using r colors.

17. Let $\{u_1, u_2, \ldots, u_5, v_1, \ldots, v_5\}$ be the vertex set of the Petersen graph P, and let the edges of P consist of those in the two 5-cycles $v_1 v_2 \ldots v_5 v_1$ and $u_1 u_3 u_5 u_2 u_4 u_1$ as well as those in the perfect matching $\{u_i v_i\}_{i=1}^{5}$. If P is in Class 1, then it has a 1-factorization E_1, E_2, E_3. We may assume that E_1 contains the edges $v_4 v_5, v_2 v_3$ and $u_1 v_1$, E_2 contains $v_1 v_2$, and E_3 contains $v_1 v_5$. This implies that $u_5 v_5 \in E_2$ and that $u_3 u_5, u_2 u_4 \in E_1$. This quickly leads to a contradiction.

Chapter 10—Domination

1. There are several thousand such placements of five queens. Using matrix location notation, placing queens at locations $(3,5)$, $(4,4), (5,4), (5,6)$, and $(6,4)$ is one way. If fewer than eight rooks are placed, then at least one row and at least one column are unoccupied. It is clear that $\gamma(K_8 \square K_8) \le 8$.

2. Any vertex that does not belong to a maximal independent set M must have a neighbor in M by the definition of "maximal" in this context.

3. This follows immediately from the fact that $D \subseteq V(G)$ is a dominating set of G if and only if $D \cap V(C_i)$ dominates C_i for each i.

4. $\lceil n/3 \rceil$ is a lower bound by Proposition 10.4. Show this bound is achieved in each of the cases $n \equiv 0 \pmod 3$, $n \equiv 1 \pmod 3$, or $n \equiv 2 \pmod 3$.

5. Consider the set $\{(1,1),(2,3),(1,5),\ldots\}$ in $P_2 \square P_n$.

6. Use Exercise 5.

7. For $v \in D$, $D - \{v\}$ does not dominate G since D is a minimal dominating set of G. Since $\deg(v) \geq 1$, it follows that $N(v) \cap (V(G) - D) \neq \emptyset$. Hence, $V(G) - D$ dominates $V(G)$.

8. The condition is equivalent to T having exactly k leaves. The set of leaves is a 2-packing, and hence $y(T) \geq k$ by Proposition 10.1, and $y(T) \leq k$ by Exercise 7. The converse can be proved by induction on k.

9. The graph H is isomorphic to $K_{4,4}$ with a perfect matching removed.

10. The largest independent set of $\overline{G_2}$ is the order of a largest complete subgraph of G_2, namely 3. Thus, $\chi(\overline{G_2}) \geq \lceil 10/3 \rceil$. Find a 4-coloring of $\overline{G_2}$. To obtain G_3, add two more triangles that share vertex b, and so on.

11. Let $M = \{u_i v_i\}_{i=1}^k$. By definition of a maximal matching, it follows that $D = \{u_1, v_1, \ldots, u_k, v_k\}$ dominates G.

12. Let $M = \{u_i v_i\}_{i=1}^k$ be a maximum matching and A be the set

$$\{u_1, v_1, \ldots, u_k, v_k\}.$$

Since M is a maximum matching in G, there does not exist $1 \leq r \leq k$ such that $x u_r, y v_r \in E(G)$ for distinct vertices x and y from $V(G) - A$ (because otherwise replacing $u_r v_r$ by $x u_r$ and $y v_r$ would give a larger matching). This implies that a dominating set of G can be formed by choosing one of u_i or v_i for each $1 \leq i \leq k$.

13. If A dominates $G-e$, then A dominates G. Hence $y(G) \leq y(G-e)$. Let $e = uv$. If there is a minimum dominating set of G that contains neither u nor v, or one that contains both of u and v, then $y(G - e) = y(G)$. Otherwise, each minimum dominating set, D, of G contains exactly one of u and v. But then $D \cup \{u, v\}$ dominates $G-e$, and thus $y(G-e) \leq y(G)+1$. If A is a minimum dominating set of $G-v$, then $A \cup \{v\}$ dominates G. Hence, $y(G) \leq y(G - v) + 1$.

14. Consider a minimum dominating set D of G. Fix an ordering $e_1, e_2, \ldots, e_m$ of the edges of G. Let $k = 1$ and $G_1 = G$. Repeat the following two steps for k, going from 1 to $m - 1$. If $G_k - e_k$ is connected and D is a dominating set of $G_k - e_k$, then let $G_{k+1} = G_k - e_k$ and increase k by 1. Otherwise, let $G_{k+1} = G_k$ and increase k by 1.

Let H be the graph G_m. By construction, H is connected and D is a dominating set of H. One can now argue that H is a tree. Indeed, if H has a cycle C, then there will be a pair of consecutive vertices on C, neither of which belongs to D. (Why must this be so?) This is a contradiction since the edge joining these two vertices would have been removed by the procedure. Thus, H is a spanning tree of G, and D is a dominating set of H.

15. For a vertex u of G with $\deg(u) = \delta(G)$, $d_G(u, v) \leq 2$ for every vertex v of G. Thus, $N(u)$ is a dominating set of G, and $y(G) \leq \deg(u) = \delta(G)$.

Chapter 11—Domination in Cartesian Products

1. Suppose there is $h \in H - B$. Then for every $g \in G$, there exists $(g, h') \in D$ such that $hh' \in E(H)$. This implies that $A = V(G)$ and B dominates H.

2. If there exist vertices $g \in G - A$ and $h \in H - B$, then (g, h) is not adjacent to any vertex of $A \times B$. This implies that $A \times B$ does not dominate $G \square H$. The converse is immediate.

3. This is a direct result of Proposition 10.4.

4. Since Q_n is n-regular, $2^n = |Q_n| = |C|(n + 1)$. Thus, $|C| = 2^r$ for some r, and $n + 1 = 2^{n-r}$. Let $k = n - r$.

5. By Proposition 11.3, $y(G \square \overline{G}) \geq n$. It is straightforward to check that $\{(g, g) \mid g \in G\}$ dominates $G \square \overline{G}$.

6. This is a direct application of Proposition 11.4 since $y(C_{3r}) = r$.

7. By Proposition 10.4, $y(C_n \square C_m) \geq \frac{nm}{5}$. Now show that, with the exception of $(n, m) = (4, 4)$ and $(n, m) = (7, 4)$, $\frac{nm}{5} \geq \lceil \frac{n}{3} \rceil \lceil \frac{m}{3} \rceil$. These two cases can be treated by using the fact that $y(C_4 \square C_4) \geq 4$ and $y(C_7 \square C_4) \geq \lceil \frac{28}{5} \rceil = 6$.

8. Let D be a dominating set of $G \square H$. Fix a vertex h of H. The only vertices of D that are adjacent to the vertices of the fiber G^h are those in $D \cap (V(G) \times N_H[h])$. This implies that $p_G(D \cap$

$(V(G) \times N_H[h]))$ dominates G, and thus $|D \cap (V(G) \times N_H[h])| \geq$ $y(G)$. It follows that $y(G \square H) \geq y(G)\rho(H)$. This implies Vizing's Conjecture is true for C_{3n}, since $y(C_{3n}) = n = \rho(C_{3n})$.

9. Let L_1 denote the set of vertices of T_1 that have degree one, and let L_2 denote the set of vertices of degree one in T_2. By the description of these trees, it is clear that $|T_1| = 2|L_1|$ and that L_1 is a 2-packing of T_1. A similar statement is true of T_2 and L_2. That T_1 and T_2 have a domination number one-half their respective orders now follows from Proposition 10.1 and Exercise 7 of Chapter 10.

For any set of vertices S in either of the trees T_1 or T_2, let S' be the set of vertices of degree one that have a neighbor in S. Let D_1 be a minimum dominating set of the subgraph of T_1 induced by $V(T_1) - L_1$ and let $E_1 = V(T_1) - (L_1 \cup D_1)$. Note that E_1 dominates this subtree of T_1 as well by Exercise 7 of Chapter 10. Similarly, we define D_2 and E_2 in the tree T_2.

The set $(D_1' \times D_2) \cup (E_1' \times E_2) \cup (D_1 \times E_2') \cup (E_1 \times D_2')$ dominates $T_1 \square T_2$ and has order $y(T_1)y(T_2)$.

10. See the solution of Exercise 9.

11. Note that $y(P_4) = 2 = y(C_4)$, and Theorem 11.8 implies that Vizing's Conjecture is true for C_4. By Proposition 10.2,

$$y(G \square P_4) \geq y(G \square C_4) \geq y(G)y(C_4) = 2y(G).$$

12. Let P_3 have vertices u, v, w, where $\deg(v) = 2$. If D is a minimum dominating set for $G \square P_3$, then $|D| \geq y(G)$ by Theorem 11.8. By considering a projection of $D \cap (G \times \{u, v\})$ into G^u, it follows that $|D \cap (V(G) \times \{u, v\})| \geq y(G)$. Similarly, $|D \cap (V(G) \times \{v, w\})| \geq y(G)$. If $|D| = y(G)$, then this forces $|D \cap G^u| = 0 = |D \cap G^w|$, and hence $D = V(G^v)$. But, this implies that G has no edges.

13. Let H be any graph. Then,

$$y(G' \square H) \geq y(G \square H) \geq y(G)y(H) = y(G')y(H).$$

The first inequality is a result of Proposition 10.2 and the second is because G satisfies Vizing's Conjecture.

Part IV—Metric Aspects

Chapter 12—Distance Lemma and Wiener Index

1. Let P be a shortest path between two vertices of the same fiber, say between (g, h) and (g', h). Then the projection $p_G P$ is a g, g'-walk in G. Hence, as P is shortest, the projection must be one-to-one, and consequently P is contained in G^h.

2. Let $e_1 = u_1 v_1, \ldots, e_m = u_m v_m$ be the edges of a tree T. For a vertex x of T, define the m-tuple $(x_1, \ldots, x_m)$ as follows: For all $i \in \{1, \ldots, m\}$ set $x_i = 0$ if $d(x, u_i) < d(x, v_i)$ and $x_i = 1$ otherwise. Consider the m-tuples as vertices of Q_m and verify that this embedding is isometric. (Use the fact that there is only one shortest path between any pair of vertices of T.)

3. Suppose G and H are partial cubes. Then G is an isometric subgraph of Q_r (for some r) and H is an isometric subgraph of Q_s (for some s). Assign to the vertex (g, h) of $G \square H$ the concatenation of the corresponding r-tuple of g and the s-tuple of h. Verify that this gives an isometric embedding of $G \square H$ into Q_{r+s}. (Use the Distance Lemma.)

4. For a graph X and vertices u, v, w of X, call a vertex from $I_X(u, v) \cap I_X(u, w) \cap I_X(v, w)$ a *median* of u, v, w.

    Let G and H be median graphs and let (g_1, h_1), (g_2, h_2), and (g_3, h_3) be arbitrary vertices of $G \square H$. As G is median, there exists a unique median g of g_1, g_2, g_3 in G, and since H is median, there is a unique median h of h_1, h_2, h_3 in H. Then verify that (g, h) is a unique median of (g_1, h_1), (g_2, h_2), and (g_3, h_3) in $G \square H$.

5. Check that in Q_n the vertices

$$000\ldots00$$
$$100\ldots00$$
$$110\ldots00$$
$$\vdots$$

$$111\ldots10$$
$$111\ldots11$$
$$011\ldots11$$
$$001\ldots11$$
$$\vdots$$
$$000\ldots00$$

induce an isometric subgraph C_{2n}. Show that this subgraph is not convex for $n > 2$.

6. Adjacent pairs of vertices in any tree T with n edges contribute n to $W(T)$. Any other pair of vertices is at distance at least two. The only tree in which each pair of nonadjacent vertices is at distance two are the stars. Therefore,

$$W(K_{1,n}) = n + 2\binom{n}{2} = n^2.$$

7. Let T be a tree that is not a path. Then T contains vertices of degree more than two. Let u be such a vertex with the additional property that at least two connected components of $T - u$ are paths. (Verify that such a vertex exists!) Select two such components and let v and w be pendant vertices (in T) on these components. We may assume that $d(u, v) \le d(u, w)$.

A typical situation is shown in Figure 4.

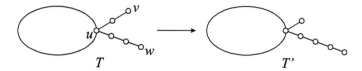

Figure 4. A tree transformation.

Let T' be the tree that is obtained from T be removing the vertex v and attaching a pendant vertex to w; see Figure 4 again. Deduce now that $W(T') > W(T)$. Proceed until the path with the same number of edges as T is obtained.

8. First determine that $W(P_n) = \binom{n+1}{3}$. Then apply Proposition 12.5 to get:

$$W(P_m \,\square\, P_n) = m^2\binom{n+1}{3} + n^2\binom{m+1}{3}.$$

9. Let H be a disconnected subgraph of a connected graph G, and let u, v be vertices that are in different components of H. Since G is connected, there exists a u, v-path in G but there is no such path in H. Thus, H cannot be convex.

Chapter 13—Products and Boxes

1. Let P a shortest path in G that connects two vertices of $p_G B$. Then $P \times h$ is a shortest path in $G \square H$ that connects two vertices of B, where $h \in p_H B$.

   If B is convex, then $P \times h$ is in B, and thus $P = p_G(P \times h)$ is in $p_G B$. Hence $p_G B$ is convex. Analogously, we can show that the convexity of B implies that of $p_H B$.

   Now let Q be a shortest path in $G \square H$ that connects two vertices of B. By the Distance Lemma, both projections $p_G Q$ and $p_H Q$ must be shortest paths in G (respectively H), that connect vertices from $p_G B$ (respectively $p_H B$). If $p_G B$ and $p_H B$ are convex, then $p_G Q$ and $p_H Q$ are in $p_G B$ (respectively $p_H B$), and Q is in B.

2. Let S be the subgraph of T that does not contain the end vertices of T. It suffices to show that T and S have the same center, if T has more than two vertices.

   Recall that w is central, if $e(w) = \text{rad}(T)$, where

   $$\text{rad}(T) = \min_{u \in T} e(u) = \min_{u \in T} \max_{v \in T} d(u, v).$$

   Except in the trees on one or two vertices, the center will clearly not be an end vertex. Thus, it suffices to consider only vertices $u \in S$ for the minimum search:

   $$\text{rad}(T) = \min_{u \in S} \max_{v \in T} d(u, v).$$

   Now we observe that $\max_{v \in T} d(u, v) = \max_{v \in S} d(u, v) + 1$ if $u \in S$. Hence, $\min_{u \in S} \max_{v \in T} d(u, v)$ will be attained by the same vertices that attain $\min_{u \in S} \max_{v \in S} d(u, v)$; that is, the central vertices of S.

3. Let T consist of the star K_{1,n^2} whose central vertex is identified with one endpoint of a path of length n. A short computation shows that the single vertex of degree $n^2 + 1$ is the distance center, whereas the center consists of one or two vertices of degree two for $n > 2$.

4. This is an immediate consequence of the Distance Lemma.

5. Let v be in the distance center of a graph G and φ an automorphism of G. Since distances are preserved under automorphisms, the image φv of v under φ must also be in the distance center of G.

6. Clearly $G \supseteq f(G)$; hence, $f(G) \supseteq f^2(G)$, and $f^j G \supseteq f^{j+1}(G)$ for all natural numbers j. Since the $f^j G$ are finite, there is one with a smallest number of vertices, say $f^k G$. But then $f^k(G) = f^{k+1}(G)$.

7. Set $H = f^k(G)$. By Exercise 6, $f(H) = H$. Therefore f is a bijection of $V(H)$. Since it also preserves adjacencies it must be an automorphism.

8. Let k be defined as in Exercise 6 and H as in Exercise 7. Since f is an automorphism and H is finite, there exists a power f^r of f that is the identity mapping of H. Then

$$f^{k(r+1)}(G) = f^{kr} f^k(G) = f^{kr}(H) = H.$$

Furthermore, since f^r is the identity mapping on H, $f^{k(r+1)}$ also is the identity mapping. Setting $g = f^{k(r+1)}$, we thus have $g^2 = g$. In other words, g is a retraction.

9. Let G be a Cartesian product and f a contraction of G. By Exercises 6, 7, and 8, there is a power f^k of f for which H is a retract of G and on which f acts as an automorphism. Since H is a retract of a Cartesian product, its distance center is a box, by Theorem 13.5. Let B be this box. By Exercise 5, automorphisms preserve distance centers; hence, $f(B) = B$.

10. Let G be the one-sided infinite path $v_0 v_1 v_2 \ldots$ and let f be the mapping $f : v_i \mapsto v_{i+1}$. Clearly, f is a contraction, but $f^i \neq f^j$ for all pairs of distinct natural numbers i, j.

Chapter 14—Canonical Metric Representation

1. For instance, xy is in relation Θ to uv as well as to all three edges in the triangle that contains v but not u. On the other hand, uv is in relation with xy, the middle vertical edge, and the two edges in the triangle containing uv.

2. In all the cases, we need only to write down the related distances. We demonstrate this for (ii). Let $P = u_0 u_1 \ldots u_m$ and $e = u_i u_{i+1}$, $f = u_j u_{j+1}$, where $i < j$. Then $d(u_i, u_j) + d(u_{i+1}, u_{j+1}) = (d(u_{i+1}, u_j) + 1) + (d(u_i, u_{j+1}) - 1)$; hence, e is not in relation Θ with f.

3. The distance between B_1 and B_2 is determined by the distance between the corresponding cut vertices. Now write down the distances between the end vertices of e_1 and e_2.

4. First solution: Let T be an arbitrary spanning tree of G containing f. If T also contains f', we are done. Otherwise, $T \cup \{f'\}$ contains exactly one cycle C. A desired spanning tree is obtained by removing any edge of C different from f'.

   Second solution: Use Kruskal's greedy algorithm and start it with the edges f and f'.

5. We can assume without loss of generality that $G = G_1 \square G_2$, $e = uv$, $f = xy$, and that u and v differ in the second coordinate. Hence, $u = (u_1, u_2)$, $v = (u_1, v_2)$, $x = (x_1, x_2)$, $y = (x_1, y_2)$. Moreover, since $p_{G_i}(e) = p_{G_i}(f)$, we may further—without loss of generality—assume that $x_2 = u_2$ and $y_2 = v_2$. Then

$$
\begin{aligned}
d(u,x) + d(v,y) &= d(u_1, x_1) + d(u_2, u_2) + d(u_1, x_1) + d(v_2, v_2) \\
&= 2d(u_1, x_1) \\
&\neq 2d(u_1, x_1) + 2 \\
&= d(u_1, x_1) + d(u_2, v_2) + d(u_1, x_1) + d(v_2, u_2) \\
&= d(u,y) + d(v,x).
\end{aligned}
$$

Hence, $e \Theta f$.

6. We can assume without loss of generality that $G = G_1 \square G_2$, $e = uv$, $f = xy$, and that u and v differ in the second coordinate. Since $e \Theta f$ we have

$$
d(u,x) + d(v,y) \neq d(u,y) + d(v,x).
$$

By the Distance Lemma, and since $u_1 = v_1$, we infer

$$
\begin{aligned}
d(u_1, x_1) \quad &+ \quad d(u_2, x_2) + d(u_1, y_1) + d(v_2, y_2) \\
&\neq \quad d(u_1, y_1) + d(u_2, y_2) + d(u_1, x_1) + d(v_2, x_2).
\end{aligned}
$$

Hence,

$$
d(u_2, x_2) + d(v_2, y_2) \neq d(u_2, y_2) + d(v_2, x_2).
$$

Thus, $x_2 \neq y_2$.

7. The main idea of the proof is to show that t (the number of factors) is equal to the number of equivalence classes of Θ^*. For

this sake, the following claim is crucial. Let U and V be arbitrary H_i-fibers, then:

Claim: $p_{G_i}(\beta(G) \cap U) \subseteq p_{G_i}(\beta(G) \cap V)$, or vice versa.

Suppose it is not. Then there are vertices $u, u' \in U$ and $v, v' \in V$ such that $u \in \beta(G) \cap U$, $u' \notin \beta(G) \cap U$, $v \in \beta(G) \cap V$, and $v' \notin \beta(G) \cap V$, where $p_{G_i}(u) = p_{G_i}(v')$ and $p_{G_i}(u') = p_{G_i}(v)$. Suppose that the distance between U and V in H is r. Because $\beta(G)$ is isometric in H, we have $r + 1 = d_H(u, v) = d_{\beta(G)}(u, v)$, which is possible only if $u' \in \beta(G)$ or $v' \in \beta(G)$. This proves the claim.

Using the claim, Exercise 5, and the fact that fibers are complete, we infer that any two edges in $\beta(G)$ that differ in the same coordinate are in relation Θ^*. However, no two edges that differ in different coordinates are in a common equivalence class, by Lemma 14.6.

Hence we have proved that the number of factors is equal to the number of equivalence classes of Θ^*. Since, by Theorem 14.7, α is unique among isometric irredundant embeddings, we conclude that $\beta = \alpha$.

8. The inequality is clear because $d_{G/F_i}(C_j^{(i)}, C_{j'}^{(i)}) \geq 1$ for any $C_j^{(i)} \neq C_{j'}^{(i)}$. Moreover, the equality holds if and only if $d_{G/F_i}(C_j^{(i)}, C_{j'}^{(i)}) = 1$ for all i and any pair of connected components $C_j^{(i)}, C_{j'}^{(i)}$. This means that all G/F_i are complete.

9. Using Exercise 7, we can design an algorithm as follows: Compute the canonical representation of a given graph G. Then G is an isometric subgraph of products of complete graphs if and only if all factors of the representation are complete.

A computationally simpler algorithm can be designed by using Exercise 8. It suffices to compute the Θ^*-classes, the components $C_j^{(i)}$, and then to check the condition of the exercise.

The recognition complexity in both cases is $O(nm)$, where n and m are the number of vertices and edges.

10. Apply Theorem 14.8, using the fact that every G/F_i is a K_2.

Part V—Algebraic and Algorithmic Issues

Chapter 15—Prime Factorizations

1. If $i = j$, then G_i^v and G_i^w are either disjoint or identical. For $i \neq j$, the case $k = 2$ has been treated in Exercise 8 of Chapter 1. If $k > 2$, there is an index $\ell \neq i, j$. Show that the intersection is empty if $p_\ell v \neq p_\ell w$, and that it consists of exactly one vertex otherwise.

2. A triangle T is prime and it is convex as a subgraph of any graph. If T is a subgraph of a Cartesian product, then by Proposition 13.3, T is a box and so T must be entirely contained in some fiber of the product. The result follows.

3. This follows directly from the Square Property (Lemma 13.1).

4. Both of G and H are connected. The assumptions about G and H imply that H contains a path h_1, h_2, h_3 and G has an edge $g_1 g_2$. The subgraph induced by $\{g_1, g_2\} \times \{h_1, h_2, h_3\}$ contains the required squares.

5. Assume that $1 \leq m \leq n$. For any n, the graph $K_{1,n}$ is prime since nontrivial Cartesian products contain no vertices of degree one. Thus, we consider the case $m \geq 2$. Suppose $K_{m,n}$ is not prime, say $K_{m,n} = G \square H$, where G and H are nontrivial and connected. Let $g_1 g_2 \in E(G)$ and $h_1 h_2 \in E(H)$. Since m of the vertices of $K_{m,n}$ have the same open neighborhood and the other n have the same open neighborhood, we may assume without loss of generality that (g_1, h_1) and (g_2, h_2) have the same open neighborhood of order n. But these two vertices are in different G fibers and in different H fibers, and as such, have only two distinct common neighbors. Thus, $m = n = 2$.

6. Suppose $u = (a, b)$. We claim that $v = (a, h)$ is the unique vertex satisfying the equation for every $x = (x', h) \in G^h$. From the Distance Lemma, we get

$$d(u,v) + d(v,x) = d_G(a,a) + d_H(b,h) + d_G(a,x') + d_H(h,h)$$
$$= d_H(b,h) + d_G(a,x')$$
$$= d(u,x).$$

If $w = (c, h) \neq v$ is any other vertex of G^h, then

$$d(u,w) + d(w,v) = d_G(a,c) + d_H(b,h) + d_G(c,a)$$
$$> d_H(b,h) = d(u,v).$$

That is, the required equation fails for $x = v \in G^h$, and hence v is unique.

7. See the solution of Exercise 6.

8. Let $G = K_4$ with vertex set $\{1, 2, 3, 4\}$. The triangle W induced by $\{1, 2, 3\}$ together with $u = 4$ is such an example.

9. Let the vertex set of the first factor be $\{1, 2, 3\}$ and that of the third factor be $\{a, b, c\}$. In addition, let $\{r, s\}, \{x, y, z\}$ be the bipartition of $K_{2,3}$. Following the specifications of the exercise, we must have $\pi(1) = 3, \pi(2) = 2$, and $\pi(3) = 1$. One example of such an automorphism φ has $\psi_1(1) = c, \psi_1(2) = a, \psi_1(3) = b$; $\psi_3(a) = 1, \psi_3(b) = 2, \psi_3(c) = 3$; and $\psi_2(r) = s, \psi_2(s) = r, \psi_2(x) = x, \psi_2(y) = z, \psi_2(z) = y$.

Now,

$$\varphi(u_1, u_2, u_3) = (\psi_{\pi^{-1}1}(u_{\pi^{-1}1}), \; \psi_{\pi^{-1}2}(u_{\pi^{-1}2}), \; \psi_{\pi^{-1}3}(u_{\pi^{-1}3}))$$
$$= (\psi_3(u_3), \; \psi_2(u_2), \; \psi_1(u_1)).$$

For example, $\varphi(3, y, a) = (\psi_3(a), \psi_2(y), \psi_1(3)) = (1, z, b)$.

10. Since the factors are pairwise nonisomorphic, connected, and prime, the permutation π in Theorem 15.5 is the identity. Therefore, the automorphisms of the factors act independently to produce all automorphisms of the product. This gives the direct product of the automorphism groups of the factors.

11. Consider two disjoint copies of the graph G, say $G_1 = G$ and $G_2 = G$, such that for $g \in G_1$, the corresponding vertex of G_2 is labeled g'. We employ Corollary 15.6. An automorphism α of $G_1 + G_2$ that does not interchange G_1 and G_2 must be the identity since G has a trivial automorphism group. But if $\alpha G_1 = G_2$, then for every $g \in G_1$ it is also the case that $\alpha g = g'$ and $\alpha g' = g$ since $\mathrm{Aut}(G)$ is trivial. Thus, $\mathrm{Aut}(G \,\square\, G)$ is cyclic of order two.

12. Each of the k factors of Q_k is K_2. Hence the permutation π of Theorem 15.5 can be any of the possible $k!$ permutations of $\{1, 2, \ldots, k\}$. In addition, each factor has an automorphism group of order two. The conclusion follows.

13. Using Corollary 15.6, it is easy to see that the symmetric group S_k is a subgroup of $\mathrm{Aut}(Q_k)$. Hence, for $k \geq 3$, $\mathrm{Aut}(Q_k)$ is non-abelian. The automorphism group of Q_2 (i.e., the cycle of order four), which is the dihedral group of order eight, is nonabelian.

14. Define $\alpha, \beta : G \,\square\, G \,\square\, G \rightarrow G \,\square\, G \,\square\, G$ by $\alpha(g_1, g_2, g_3) = (g_2, g_1, g_3)$ and $\beta(g_1, g_2, g_3) = (g_3, g_2, g_1)$. Then, $\alpha, \beta \in \operatorname{Aut}(G \,\square\, G \,\square\, G)$, but $\alpha\beta \neq \beta\alpha$.

15. Choose any nonidentity element σ of $\operatorname{Aut}(G)$. Assume $\sigma x \neq x$. Define $\alpha, \beta : G \,\square\, G \rightarrow G \,\square\, G$ by $\alpha(g_1, g_2) = (\sigma g_1, g_2)$ and $\beta(g_1, g_2) = (g_2, g_1)$. It is now easy to show that $\alpha, \beta \in \operatorname{Aut}(G \,\square\, G)$ and that $\alpha\beta(x, \sigma^{-1}x) \neq \beta\alpha(x, \sigma^{-1}x)$.

16. If neither n nor m is 1, then Proposition 1.2 and the surrounding work show $L(K_{m,n}) = K_m \,\square\, K_n$. Since $\operatorname{Aut}(K_r) = S_r$ for any r, the result follows by an application of Exercise 10.

Chapter 16—Cancelation and Containment

1. Let $\sigma : G \,\square\, (H + K) \rightarrow G \,\square\, H + G \,\square\, K$ be the function defined by $\sigma(g, x) = (g, x)$. Verify that σ is a graph isomorphism. A similar natural function will give an isomorphism from $(G + H) \,\square\, K$ to $G \,\square\, K + H \,\square\, K$.

2. Since G, H, and K are connected, Theorem 15.1 implies there exist prime graphs $R_1, R_2, \ldots, R_k, S_1, S_2, \ldots, S_m$ and $T_1, T_2, \ldots, T_n$ such that $G = \square_{i=1}^{k} R_i$, $H = \square_{i=1}^{m} S_i$, and $K = \square_{i=1}^{n} T_i$. (Repetitions of prime factors are allowed in this notation.) Theorem 15.1 also guarantees a unique prime factorization of the connected graph $G \,\square\, K = H \,\square\, K$. By combining the above factorizations of G, H, and K, we see that

$$G \,\square\, K \;=\; R_1 \,\square\, R_2 \,\square\, \cdots \,\square\, R_k \,\square\, T_1 \,\square\, T_2 \,\square\, \cdots \,\square\, T_n$$
$$H \,\square\, K \;=\; S_1 \,\square\, S_2 \,\square\, \cdots \,\square\, S_m \,\square\, T_1 \,\square\, T_2 \,\square\, \cdots \,\square\, T_n.$$

By the uniqueness of the prime factorization of $G \,\square\, K = H \,\square\, K$, it now follows that $G = \square_{i=1}^{k} R_i = \square_{i=1}^{m} S_i = H$.

3. First, assume that G and H are connected. Equation (16.3) is true in this case because of Theorem 15.1 and the definition of P, and Equation (16.4) is how P is defined. For the case in which G or H is disconnected, we apply Theorem 15.1 on each component, use the distributive property in $\mathbb{S}(\Gamma)$, and use the reduction (to connected graphs) above.

4. It suffices to show that the factors on the left-hand and right-hand sides are prime. Both of $K_1 + K_2 + K_2^2$ and $K_1 + K_2$ are prime since they have prime orders. Also, the number of components of a Cartesian product is the product of the number of components

in the factors. Hence, exactly one prime factor of $K_1 + K_2^2 + K_2^4$ has three components, and the others have just one. Clearly, every other factor has to divide all components of $K_1 + K_2^2 + K_2^4$, and thus each must be K_1. Therefore, $K_1 + K_2^2 + K_2^4$ is prime. A similar argument shows that $K_1 + K_2^3$ is prime.

5. Let K_2 and K_3 correspond to x_1 and x_2, respectively. If $G = C_4 + K_2^2 \square K_3 + 2(K_2 \square K_3) + K_2 \square K_3^2 + K_3^2$, then $P(G) = x_1^2 + x_1^2 x_2 + 2x_1 x_2 + x_1 x_2^2 + x_2^2$. Now show that the unique prime factorization of $P(G)$ in the ring $\mathbb{Z}[X]$ is $P(G) = (x_1 + x_2)(x_1 + x_1 x_2 + x_2)$, and then "read off" the prime graphs arising from these polynomial factors.

6. Every edge of C_4^2 belongs to three chordless 4-cycles, but any edge of C_6^2 is in only two such 4-cycles.

7. One way to do it is to connect an arbitrary vertex of one connected component $K_3 \square K_3$ with an arbitrary vertex of the other component $P_4 \square P_4$ of H. Verify that the required conditions still hold!

8. Use the graphs G and H from Exercise 7. To make G prime, remove one of its edges. The graph H constructed in the solution of Exercise 7 is already prime. Again check that the required conditions still hold!

9. Let G be a disconnected, vertex-transitive graph with a prime factorization $G_1 \square G_2 \square \cdots \square G_k$. All components are isomorphic to a graph H (which is also vertex transitive). Let $H_1 \square H_2 \square \cdots \square H_\ell$ be the prime factorization of H. It is easy to see that every component of a G_i must be a subproduct of $H_1 \square H_2 \square \cdots \square H_\ell$.

   Show that the components of G cannot be isomorphic if this is not the case for the components of G_i.

Chapter 17—Distinguishing Number

1. We prove that $D(C_n) = 3$. Clearly $D(C_n)$ cannot be 1 because C_n is not asymmetric. Also, if we choose two adjacent vertices, label one of them blue, the other black, and the rest white, then we obtain a distinguishing 3-labeling.

   Suppose now $D(C_n) = 2$ for $n = 3, 4, 5$. We have to show the existence of a nontrivial, label-preserving automorphism. If one vertex is black and the others white, then there exists a nontrivial automorphism of order two that fixes the black vertex. If there are exactly two black vertices, then they are either adjacent or of distance two. In both cases, there exists an automorphism that interchanges them.

Finally, the case of three or more black vertices and at least one white vertex can be reduced to one of the already treated cases by interchanging the labels.

2. There is just one black vertex with three white neighbors; it must be fixed by any label-preserving automorphism. But then its only black neighbor must also be fixed, and thus also the third black vertex.

Furthermore, the only white vertex with no black neighbor must be fixed, but then also the unique neighbors of distances four, three, and two from this vertex. This leaves two white vertices that are not fixed yet. One of them has two black neighbors, the other just one. So they must also be fixed.

3. It suffices to show that two elements u, v of S are adjacent if and only if they are adjacent in the lexicographic ordering of their coordinate vectors. We set

$$S_j = \{(j+1, j, \ldots, j), (j+1, j+1, j, \ldots, j), \ldots, (j+1, j+1, \ldots, j+1)\}$$

for $j = 1, \ldots, n - 1$. Clearly, the S_j are mutually disjoint and

$$S = \{(1, 1, \ldots, 1)\} \cup \bigcup_{j=1}^{n-1} S_j.$$

Let u, v be in S_j. Both u and v must have indices valued $j + 1$. If the number of indices with value $j + 1$ is the same, then $u = v$. If the numbers differ by one, then u and v are adjacent; otherwise they are nonadjacent. Thus, two vertices of S_j are adjacent if and only if they are adjacent in the lexicographic ordering of their coordinate vectors.

Let $u \in S_j$ and $v \in S_{j'}$, where $j < j'$. Suppose $j + 2 \le j'$. By definition, u has no coordinates valued $j + 2$, whereas all coordinates of $S_{j'}$ have the values $j + 2$ or larger. Hence, u and v are nonadjacent unless $j' = j + 1$, that is, $v \in S_{j+1}$. But then all coordinate values of v are $j + 1$ or $j + 2$, but not j. Since there is only one vertex in S_j that has no coordinate valued j, namely the vertex $(j + 1, j + 1, \ldots, j + 1)$, this vertex must be u. Furthermore, the only element of S_{j+1} that is adjacent to this vertex is

$(j + 2, j + 1, \ldots, j + 1)$; it must be v. Clearly u and v follow each other in the lexicographic ordering of the coordinate vectors of S.

The proof is completed by the observation that the only neighbor of $(1, 1, \ldots, 1)$ in S is $(2, 1, \ldots, 1)$.

4. Let G be $K_3 \square K_3$ or Q_3, and let S be the induced path from the proof of Proposition 17.5. We label the first vertex of S blue, all other vertices of S black, and the rest white.

Every label-preserving automorphism φ of G must preserve S and thus induce a label-preserving automorphism on S. Since the only such automorphism fixes S pointwise, φ must be the identity by Proposition 17.5. Hence our labeling is distinguishing.

5. By Exercise 4, it remains to show that $K_3 \square K_3$ has no distinguishing 2-labeling. We show that every 2-labeling admits a nontrivial label-preserving automorphism.

Every 2-labeling has a set of labels with at most four vertices; it suffices to consider the cases where exactly one, two, three, and four vertices are labeled black, and the others white.

(a) Suppose we have a 2-labeling with exactly one black label. Let $(1, 1)$ be this vertex. Clearly an interchange of the factors, that is, the automorphism $(x, y) \mapsto (y, x)$, is a nontrivial label-preserving automorphism.

(b) If there are exactly two black vertices, they can be adjacent or nonadjacent. In the first case, we can assume these vertices are $(1, 1)$ and $(2, 1)$. Then the mapping, say β, that interchanges the vertices $(1, y)$ and $(2, y)$ (for $y = 1, 2, 3$) and fixes the others is a nontrivial label-preserving automorphism.

Now assume that the black vertices are $(1, 1)$ and $(2, 2)$. Again, an interchange of the factors provides a nontrivial label-preserving automorphism.

The cases (c) and (d) with three and, respectively, four black vertices are treated similarly.

6. Similar to the case $K_3 \square K_3$, it remains to show by Exercise 4 that Q_3 has no distinguishing 2-labeling. We show that every 2-labeling admits a nontrivial label-preserving automorphism.

Since Q_3 has eight vertices, it suffices to consider labelings of exactly one, two, three, or four black vertices, where all other vertices are white.

(a) There is just one black vertex, say $(1, 1, 1)$. Then an interchange of any two factors is the desired automorphism.

(b) There are exactly two black vertices u, v. Then they can have distance one, two, or three. In the first case, they are in one and the same fiber with respect to a factor, say the first one. In this case the (only) nontrivial automorphism of that factor induces the desired automorphism.

If they have distance two, then there is a factor such that the projections of u, v into this factor are the same, whereas the projections into any one of the other two factors are distinct. We interchange these two factors.

If u, v have distance three, then we can assume that they are the antipodal vertices $(1, 1, 1)$ and $(2, 2, 2)$. In this case we can interchange any two factors.

As in Exercise 5, we leave cases (c) and (d) with three and four, respectively, black vertices to the reader.

7. Suppose G and K_n are not relatively prime. Since K_n is prime, it must be a factor of G. But then G cannot be asymmetric.

8. By Exercise 7, G and K_n are relatively prime. This implies that all the automorphisms of $G \square K_n$ are induced by those of the factors. Since G is asymmetric, this means that all automorphisms of

$$G \square K_n$$

are induced by those of K_n. In other words, every automorphism of $G \square K_n$ permutes the G-fibers without changing them, and every permutation of the G-fibers is an automorphism.

In order to find a distinguishing labeling it suffices to label G in n different ways. Since G has k vertices, one can label G in d^k ways with d labels. Thus the smallest d with $d^k \geq n$ is the distinguishing number of $G \square K_n$.

9. Let U, W be the bipartition of $K_{n,n}$. Then the permutation of $U \cup W$ that interchanges any two vertices of U (or any two vertices of W) and fixes all other vertices is an automorphism. Thus, a labeling that assigns the same labels to two distinct vertices of U or W is not distinguishing. We thus need n distinct labels for U and n for W. They cannot be the same; otherwise, a label-preserving interchange of U and W would be possible.

10. Recall from Chapter 1 (Proposition 1.2) that the line graph of $K_{m,n}$ is $K_m \square K_n$. Thus, every edge coloring of $K_{m,n}$ is a vertex labeling of $K_m \square K_n$ and vice versa.

    By the definition of the line graph, every automorphism of $K_m \square K_n$ induces one of $K_{m,n}$. Clearly, the number of automorphisms for both graphs is $m!n!$ if $m \neq n$, and $2(m!)^2$ otherwise. Thus every automorphism of $K_{m,n}$ is obtained from one of $K_m \square K_n$ and every vertex-distinguishing labeling of $K_n \square K_m$ is an edge distinguishing coloring of $K_{m,n}$. Thus $\ell(m,n) = D(K_{m,n})$.

11. Label one copy of G in $G + G$ with a $D(G)$-distinguishing labeling ℓ and the other copy with ℓ', where $\ell'(x) = \ell(x) + 1$ for any $x \in G$. This is a $(D(G) + 1)$-distinguishing labeling of $G + G$.

    Suppose G is not uniquely distinguishable. Then there exist $D(G)$-distinguishing labelings ℓ_1 and ℓ_2 of G such that no automorphism of G maps one onto the other. Labeling one copy of $G + G$ with ℓ_1 and the other copy with ℓ_2 gives a $D(G)$-distinguishing labeling of $G + G$. Conversely, suppose that $D(G + G) = D(G)$. Label $G + G$ with a $D(G)$-distinguishing labeling ℓ. Let ℓ_1 be the restriction of ℓ to one copy of G in $G + G$ and ℓ_2 the restriction of ℓ to the other copy. Then no automorphism of G maps ℓ_1 onto ℓ_2 (such that the labels are preserved); hence, G is not uniquely distinguishable.

12. We first observe that in both cases a 2-distinguishable labeling necessarily uses label 1 three times and label 2 three times. For C_6, the only way to do this (up to automorphisms) is as shown in Figure 17.1. The unique way (again up to automorphisms) for a 2-distinguishable labeling of $K_2 \square K_3$ is to label one K_3-fiber $1, 1, 2$ and the other fiber $2, 1, 2$, respectively.

Chapter 18—Recognizing Products and Partial Cubes

1. Consider the edges uv and vw. Since bipartite graphs contain no triangles, we infer that $d(u,w) = 2$. Thus, $d(u,v) + d(v,w) = 1 + 1 = 2$, and $d(u,w) + d(v,v) = d(u,w) = 2$, which implies that uv and vw are not in the relation Θ.

2. In a connected graph, the distances of two adjacent vertices x, y from any vertex v can differ by at most one. On the other hand, in a bipartite graph they cannot be equal; otherwise, there would exist a closed walk of odd length consisting of shortest paths from v to x and y and the edge xy. Thus, $d(u,x) \pm 1 = d(u,y)$ and $d(v,x) \pm 1 = d(v,y)$.

By our assumption, we infer $d(u,x)+1 = d(u,y)$. If $d(v,x)+1 = d(v,y)$, then $d(u,x) + d(v,y) = d(v,x) + d(u,y)$, contrary to our assumption.

We then have

$$1 + d(u,x) = d(v,u) + d(u,x) \leq d(v,x) = d(v,y) + 1,$$

and similarly,

$$1 + d(v,y) = d(u,v) + d(v,y) \leq d(u,v) = d(u,x) + 1.$$

Hence, $d(u,x) = d(v,y)$.

3. Let T be a spanning tree of the connected graph G, uv an arbitrary edge of $G - T$, and W the unique path from u to v in T. By Lemma 14.2, the edge uv is in relation Θ with at least one edge f of W. Thus, there cannot be more Θ^*-classes than edges of T; that is, there are at most $|G| - 1$ Θ^*-classes.

4. Let e have the endpoints u and v. It suffices to show that every walk W from u to v in G contains an edge from Θ_e. But this is clearly so by Lemma 14.2.

5. Let a, b be two vertices of F and P an a, b-path in F. We have to show that every shortest a, b-path Q of G is in F.

   Let uv be an edge of Q. If uv is not already in F, then $P \cup Q - uv$ is a u, v-walk that does not contain uv. By Lemma 14.2, there is an edge f in $P \cup Q - uv$ that is in the relation Θ to uv.

   By Lemma 14.1, no two edges of Q are in the relation Θ because Q is a shortest path. Thus, f cannot be in Q and must therefore be in P. But then uv is in relation Θ with an edge of F. Since F is closed with respect to Θ, uv is also in F.

6. Suppose the edges uv and vw of H are in the relation $\tau(H)$. Then u and w are nonadjacent and v is the only common neighbor of u and w in H. Since H is convex, uw cannot be an edge in G. Furthermore, if there are common neighbors of u and w in G, then they must already be in H by convexity. Thus, uv and vw are also in the relation $\tau(G)$.

   On the other hand, if uv and vw are in the relation $\tau(G)$, then they are clearly also in the relation $\tau(H)$, because H is convex and thus induced.

7. We first embed G_i into G such that it coincides with H. Then p_i projects G onto H.

Since H is convex, the restriction of $\Theta(G)$ to H is $\Theta(H)$.

Suppose e, f are edges of H that are in the relation $\Theta^*(G)$. Then there exists a sequence $e = e_1, e_2, \ldots, e_r = f$ of edges with $e_j \Theta(G) e_{j+1}$ for $j = 1, \ldots, r - 1$. By Lemma 14.6, all e_j are in G_i-fibers. It suffices to show that $p_i(e_j) \Theta(G) p_i(e_{j+1})$. To show this, we set

$$G_i^* = G_1 \,\square\, \cdots \,\square\, G_{i-1} \,\square\, G_{i+1} \,\square\, \cdots \,\square\, G_k$$

and note that $G = G_i \,\square\, G_i^*$. Furthermore, let $uv = e_j$ and $xy = e_{j+1}$. Since $e_j \Theta(G) e_{j+1}$, we have $d(u,x) + d(v,y) \neq d(v,x) + d(u,y)$, or, equivalently,

$$d(u,x) - d(v,x) \neq d(u,y) - d(v,y).$$

By the Distance Lemma

$$d(u,x) = d_{G_i}(u,x) + d_{G_i^*}(u,x)$$

and

$$d(u,y) = d_{G_i}(u,y) + d_{G_i^*}(u,y).$$

Since xy is in a G_i-fiber, $d_{G_i^*}(u,x) = d_{G_i^*}(u,y)$. Thus,

$$d(u,x) - d(v,x) = d_{G_i}(u,x) - d_{G_i}(u,y).$$

Similarly,

$$d(u,y) - d(v,y) = d_{G_i}(u,y) - d_{G_i}(v,y).$$

Hence, $p_i(e_j) \Theta(G_i) p_i(e_{j+1})$, which had to be shown.

8. By definition, H is a connected component of a $\sigma(G)$-class, since $(\tau \cup \Theta)^* = (\tau \cup \Theta^*)^*$. By Exercises 6 and 7, the relations $\tau(G)$ and $\Theta^*(G)$ coincide with $\tau(H)$ and $\Theta^*(H)$ on H. Hence $E(H)$ consists of only one $\sigma(H)$-class.

9. Let $v \in \square_{i=1}^k G_i$. To any two indices i, j, with $1 \leq i < j \leq k$, there are neighbors u (respectively w) of v in G that differ only in the ith (respectively the jth) coordinate from v. By the Square Property 13.1 there exists a unique vertex x in G that is adjacent to to both u and w. Applying the Square Property again to u and its neighbors v, x we see that w is the only vertex besides u that is adjacent to both v and x. This implies that the vertices u

and w have exactly one common neighbor in $G - v$, and that the edges ux and xw are in the relation $\tau(G - v)$.

We further note that removal of vertices from G that are not in a given G_i-fiber H does not alter $\sigma(H)$. Thus, H will be completely contained in a $\sigma(H)$-class. By the above reasoning, the fibers through ux and uv, that is G_i^u and G_j^u, will be in the same $\sigma(H)$-class, and thus also all other G_i- and G_j-fibers that do not contain v. Since i and j were arbitrarily chosen, this implies that all fibers of G that do not contain v are in one and the same $\sigma(H)$-class.

Finally, consider an edge e that does not contain v in G_i^v. By the Square Property it is in relation Θ to some edge in a G_i-fiber that does not contain v. Thus, e is also in the $\sigma(H)$-class of the G_i-fibers that do not contain v.

In other words, $G - v$ has only one σ-class and is prime.

10. For $G = K_{2,3}$ compare the remarks to Figure 14.2. Taking recourse to the figure, suppose G also contains the edge xv. Then it is easy to see that Θ is transitive, regardless of the dotted (...) edges, because any two edges that are not in one and the same triangle already can be connected by a chain of triangles $T_1, T_2, \ldots, T_k$, where $T_i \cap T_{i+1}$ is an edge.

If uy is an edge, then the square $xyuv$ is in one Θ^*-class. Clearly, the edges xz and vz also belong to this class.

If uz is an edge, we argue similarly, and if both uy and uz are edges, then we can again connect any two edges that are not in one and the same triangle by a chain of triangles $T_1, T_2, \ldots, T_k$, where $T_i \cap T_{i+1}$ is an edge.

11. Suppose that G has at least five vertices. Consider two adjacent G-fibers G^a and G^b. Choose a vertex (u, a) in G^a and two vertices $(x, b), (y, b)$ that are adjacent to (u, a) in $\overline{G}$. Clearly $x, y \neq u$. Since G has at least five vertices, there are two additional vertices v, w in G. Clearly, $(u, a), (v, a), (w, a), (x, b), (y, b)$ span a $K_{2,3}$ and all its edges are in one Θ^*-class by Exercise 10, regardless of the existence of other edges. Note that the edges $(u, a)(x, b)$ and $(u, a)(y, b)$ were arbitrarily chosen.

Consider any two edges e, f in $\overline{G}$ between $V(G^a)$ and $V(G^b)$. The projection of the endpoints of e and f into G comprise at most four vertices. Let w be a fifth vertex. Join (w, a) with the endpoints of e and f in G^b to form a path $ee'f'f$. Clearly, this implies $e\Theta^*f$.

Hence, all edges between $V(G^a)$ and $V(G^b)$ in $\overline{G}$ are in the relation Θ^*.

Now, consider a third fiber G^c. Consider the 4-cycle (u, a), (v, b), (w, a), (v, c) in $\overline{G}$. Regardless of the existence of diagonals, all edges of the 4-cycle clearly are in one and the same Θ^*-class. Hence, all edges in $\overline{G}$ between different G-fibers are in the same Θ^*-class.

Finally, let us consider an arbitrary edge with both endpoints in a G-fiber, say $(u, a)(v, a)$. Then there is a $w \neq u, v$ in G and a $b \neq a$ in H such that $(u, a)(w, b)(v, a)$ is a triangle in $\overline{G}$. But then all edges of $\overline{G}$ are in the same Θ^*-class and $\overline{G}$ is prime.

12. The statement is true for $k = 1$; hence, we can assume that $k > 1$. Consider an arbitrary edge uv in Q_k. We can represent Q_k as a product $K_2 \,\square\, Q_{k-1}$ such that uv is a K_2-fiber. By Lemma 14.6, no K_2-fiber can be in relation Θ to an edge in a Q_{k-1}-fiber. Thus, the Θ^*-class of uv must consist of K_2-fibers.

Furthermore, since the K_2-fibers induce an isomorphism between the two Q_{k-1}-fibers of Q_k, it is clear that any two K_2-fibers are in the relation Θ. Hence the Θ^*-class of uv must consist of the set of all K_2-fibers, and Θ is transitive.

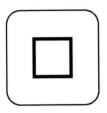

References

[1] J. ADAMSSON AND R. B. RICHTER, *Arrangements, circular arrangements and the crossing number of $C_7 \times C_n$*, J. Combin. Theory Ser. B, 90 (2004), pp. 21-39.

[2] M. O. ALBERTSON AND K. L. COLLINS, *Symmetry breaking in graphs*, Electron. J. Combin., 3 (1996). #R18, 17 pp.

[3] N. ALON, *Transversal numbers of uniform hypergraphs*, Graphs Combin., 6 (1990), pp. 1-4.

[4] K. APPEL AND W. HAKEN, *Every planar map is four colorable*, Bull. Amer. Math. Soc., 82 (1976), pp. 711-712.

[5] V. I. ARNAUTOV, *Estimation of the exterior stability number of a graph by means of the minimal degree of the vertices*, Prikl. Mat. i Programmirovanie, (1974), pp. 3-8.

[6] A. M. BARCALKIN AND L. F. GERMAN, *The external stability number of the Cartesian product of graphs*, Bul. Akad. Shtiintse RSS Moldoven., (1979), pp. 5-8, 94.

[7] D. BARNETTE, *Trees in polyhedral graphs*, Canad. J. Math., 18 (1966), pp. 731-736.

[8] V. BATAGELJ AND T. PISANSKI, *Hamiltonian cycles in the Cartesian product of a tree and a cycle*, Discrete Math., 38 (1982), pp. 311-312.

[9] M. BEHZAD AND S. E. MAHMOODIAN, *On topological invariants of the product of graphs*, Canad. Math. Bull., 12 (1969), pp. 157-166.

[10] C. BERGE, *Graphs and Hypergraphs*, North-Holland Publishing Co., Amsterdam, 1973. Translated from the French by Edward Minieka, North-Holland Mathematical Library, Vol. 6.

[11] D. P. BIEBIGHAUSER AND M. N. ELLINGHAM, *Prism-hamiltonicity of triangulations*, J. Graph Theory, 57 (2008), pp. 181-197.

[12] D. BOKAL, *On the crossing numbers of Cartesian products with paths*, J. Combin. Theory Ser. B, 97 (2007), pp. 381–384.

[13] D. BOKAL, *On the crossing numbers of Cartesian products with trees*, J. Graph Theory, 56 (2007), pp. 287–300.

[14] M. BOROWIECKI, I. BROERE, M. FRICK, P. MIHÓK, AND G. SEMANIŠIN, *A survey of hereditary properties of graphs*, Discuss. Math. Graph Theory, 17 (1997), pp. 5–50.

[15] M. BOROWIECKI, S. JENDROL, D. KRÁL, AND J. MIŠKUF, *List coloring of Cartesian products of graphs*, Discrete Math., 306 (2006), pp. 1955–1958.

[16] B. BREŠAR, *On subgraphs of Cartesian product graphs and S-primeness*, Discrete Math., 282 (2004), pp. 43–52.

[17] B. BREŠAR AND B. ZMAZEK, *On the independence graph of a graph*, Discrete Math., 272 (2003), pp. 263–268.

[18] R. L. BROOKS, *On colouring the nodes of a network*, Proc. Cambridge Philos. Soc., 37 (1941), pp. 194–197.

[19] T. CALAMONERI, *The l(h, k)-labelling problem: A survey and annotated bibliography*, Computer J., 49 (2006), pp. 585–608.

[20] G. CHARTRAND AND F. HARARY, *Planar permutation graphs*, Ann. Inst. Henri Poincaré, Nouv. Sér., Sect. B, 3 (1967), pp. 433–438.

[21] X. CHEN AND J. SHEN, *On the Frame-Stewart conjecture about the Towers of Hanoi*, SIAM J. Comput., 33 (2004), pp. 584–589.

[22] W. E. CLARK AND S. SUEN, *An inequality related to Vizing's conjecture*, Electron. J. Combin., 7 (2000). #N4, 3 pp.

[23] E. COCKAYNE, S. E. GOODMAN, AND S. T. HEDETNIEMI, *A linear algorithm for the domination number of a tree*, Inform. Process. Lett., 4 (1975), pp. 41–44.

[24] K. L. COLLINS AND A. N. TRENK, *The distinguishing chromatic number*, Electron. J. Combin., 13 (2006). #R16, 19 pp.

[25] T. H. CORMEN, C. E. LEISERSON, R. L. RIVEST, AND C. STEIN, *Introduction to Algorithms*, MIT Press, Cambridge, MA, second ed., 2001.

[26] M. M. DEZA AND M. LAURENT, *Geometry of Cuts and Metrics*, vol. 15 of Algorithms and Combinatorics, Springer-Verlag, Berlin, 1997.

[27] V. V. DIMAKOPOULOS, L. PALIOS, AND A. S. POULAKIDAS, *On the Hamiltonicity of the Cartesian product*, Inform. Process. Lett., 96 (2005), pp. 49–53.

[28] G. Dirac, *Some theorems on abstract graphs*, Proc. London Math. Soc., 2 (1952), pp. 68-81.

[29] D. Ž. Djoković, *Distance-preserving subgraphs of hypercubes*, J. Combinatorial Theory Ser. B, 14 (1973), pp. 263-267.

[30] M. El-Zahar and C. M. Pareek, *Domination number of products of graphs*, Ars Combin., 31 (1991), pp. 223-227.

[31] D. Eppstein, *Recognizing partial cubes in quadratic time.* arXiv:0705.1025v1.

[32] P. Erdős, A. L. Rubin, and H. Taylor, *Choosability in graphs*, in Proceedings of the West Coast Conference on Combinatorics, Graph Theory and Computing (Humboldt State Univ., Arcata, Calif., 1979), Congress. Numer., XXVI, Winnipeg, Man., 1980, Utilitas Math., pp. 125-157.

[33] T. Feder, *Stable Networks and Product Graphs*, Mem. Amer. Math. Soc., 116 (1995), pp. xii+223.

[34] J. Feigenbaum, J. Hershberger, and A. A. Schäffer, *A polynomial time algorithm for finding the prime factors of Cartesian-product graphs*, Discrete Appl. Math., 12 (1985), pp. 123-138.

[35] A. Fernández, T. Leighton, and J. L. López-Presa, *Containment properties of product and power graphs*, Discrete Appl. Math., 155 (2007), pp. 300-311.

[36] A. Finbow, B. Hartnell, and R. J. Nowakowski, *A characterization of well covered graphs of girth 5 or greater*, J. Combin. Theory Ser. B, 57 (1993), pp. 44-68.

[37] J. F. Fink, M. S. Jacobson, L. F. Kinch, and J. Roberts, *On graphs having domination number half their order*, Period. Math. Hungar., 16 (1985), pp. 287-293.

[38] M. J. Fisher and G. Isaak, *Distinguishing colorings of Cartesian products of complete graphs*, Discrete Math., 308 (2008), pp. 2240-2246.

[39] M. R. Garey and D. S. Johnson, *Computers and Intractability*, W. H. Freeman and Co., San Francisco, Calif., 1979.

[40] J. P. Georges and D. W. Mauro, *Some results on λ_k^j-numbers of the products of complete graphs*, in Proceedings of the Thirtieth Southeastern International Conference on Combinatorics, Graph Theory, and Computing (Boca Raton, FL, 1999), vol. 140, 1999, pp. 141-160.

[41] J. P. Georges, D. W. Mauro, and M. I. Stein, *Labeling products of complete graphs with a condition at distance two*, SIAM J. Discrete Math., 14 (2001), pp. 28-35.

[42] J. P. GEORGES, D. W. MAURO, AND M. A. WHITTLESEY, *Relating path coverings to vertex labellings with a condition at distance two*, Discrete Math., 135 (1994), pp. 103-111.

[43] L. Y. GLEBSKY AND G. SALAZAR, *The crossing number of $C_m \times C_n$ is as conjectured for $n \geq m(m+1)$*, J. Graph Theory, 47 (2004), pp. 53-72.

[44] W. GODDARD AND M. HENNING, personal communication, (2003), pp. 1-2.

[45] R. L. GRAHAM, *On primitive graphs and optimal vertex assignments*, Ann. New York Acad. Sci., 175 (1970), pp. 170-186.

[46] R. L. GRAHAM AND P. M. WINKLER, *On isometric embeddings of graphs*, Trans. Amer. Math. Soc., 288 (1985), pp. 527-536.

[47] A. GRAOVAC AND T. PISANSKI, *On the Wiener index of a graph*, J. Math. Chem., 8 (1991), pp. 53-62.

[48] J. R. GRIGGS AND R. K. YEH, *Labelling graphs with a condition at distance 2*, SIAM J. Discrete Math., 5 (1992), pp. 586-595.

[49] J. L. GROSS AND J. YELLEN, *Graph Theory and its Applications*, Discrete Mathematics and its Applications (Boca Raton), Chapman & Hall/CRC, Boca Raton, FL, second ed., 2006.

[50] J. HAGAUER AND S. KLAVŽAR, *On independence numbers of the Cartesian product of graphs*, Ars Combin., 43 (1996), pp. 149-157.

[51] F. HARARY AND L. H. HSU, *Conditional chromatic number and graph operations*, Bull. Inst. Math. Acad. Sinica, 19 (1991), pp. 125-134.

[52] F. HARARY, P. C. KAINEN, AND A. J. SCHWENK, *Toroidal graphs with arbitrarily high crossing numbers*, Nanta Math., 6 (1973), pp. 58-67.

[53] F. HARARY AND M. LIVINGSTON, *Independent domination in hypercubes*, Appl. Math. Lett., 6 (1993), pp. 27-28.

[54] B. L. HARTNELL AND D. F. RALL, *Vizing's conjecture and the one-half argument*, Discuss. Math. Graph Theory, 15 (1995), pp. 205-216.

[55] B. L. HARTNELL AND D. F. RALL, *Lower bounds for dominating Cartesian products*, J. Combin. Math. Combin. Comput., 31 (1999), pp. 219-226.

[56] B. L. HARTNELL AND D. F. RALL, *Improving some bounds for dominating Cartesian products*, Discuss. Math. Graph Theory, 23 (2003), pp. 261-272.

[57] T. HAYNES, S. HEDETNIEMI, AND P. SLATER, *Domination in Graphs*, vol. 209 of Monographs and Textbooks in Pure and Applied Mathematics, Marcel Dekker Inc., New York, 1998.

[58] S. HEDETNIEMI, *Homomorphisms of graphs and automata*, Doctoral Disseration, University of Michigan, (1966).

[59] P. HELL, X. YU, AND H. S. ZHOU, *Independence ratios of graph powers*, Discrete Math., 127 (1994), pp. 213-220.

[60] P. E. HIMELWRIGHT AND J. E. WILLIAMSON, *On 1-factorability and edge-colorability of Cartesian products of graphs*, Elem. Math., 29 (1974), pp. 66-67.

[61] A. M. HINZ, *The Tower of Hanoi*, Enseign. Math. (2), 35 (1989), pp. 289-321.

[62] W. IMRICH, *Automorphismen und das kartesische Produkt von Graphen*, Österreich. Akad. Wiss. Math.-Natur. Kl. S.-B. II, 177 (1969), pp. 203-214.

[63] W. IMRICH, *On products of graphs and regular groups*, Israel J. Math., 11 (1972), pp. 258-264.

[64] W. IMRICH, J. JEREBIC, AND S. KLAVŽAR, *Distinguishing Cartesian product graphs*, European J. Combin., 29 (2008), pp. 922-929.

[65] W. IMRICH AND S. KLAVŽAR, *A simple $O(mn)$ algorithm for recognizing Hamming graphs*, Bull. Inst. Combin. Appl., 9 (1993), pp. 45-56.

[66] W. IMRICH AND S. KLAVŽAR, *Product Graphs*, Wiley-Interscience Series in Discrete Mathematics and Optimization, Wiley-Interscience, New York, 2000.

[67] W. IMRICH AND S. KLAVŽAR, *Distinguishing Cartesian powers of graphs*, J. Graph Theory, 53 (2006), pp. 250-260.

[68] W. IMRICH, S. KLAVŽAR, AND D. F. RALL, *Cancellation properties of products of graphs.*, Discrete Appl. Math., 155 (2007), pp. 2362-2364.

[69] W. IMRICH AND I. PETERIN, *Recognizing Cartesian products in linear time*, Discrete Math., 307 (2007), pp. 472-483.

[70] M. S. JACOBSON AND L. F. KINCH, *On the domination number of products of graphs: I*, Ars Combin., 18 (1984), pp. 33-44.

[71] M. S. JACOBSON AND L. F. KINCH, *On the domination of the products of graphs II: trees*, J. Graph Theory, 10 (1986), pp. 97-106.

[72] S. JENDROL' AND M. ŠČERBOVÁ, *On the crossing numbers of $S_m \times P_n$ and $S_m \times C_n$*, Časopis Pěst. Mat., 107 (1982), pp. 225-230, 307.

[73] P. K. JHA, S. KLAVŽAR, AND A. VESEL, *Optimal $L(d, 1)$-labelings of certain direct products of cycles and Cartesian products of cycles*, Discrete Appl. Math., 152 (2005), pp. 257-265.

[74] P. K. JHA AND G. SLUTZKI, *A note on outerplanarity of product graphs*, Zastos. Mat., 21 (1993), pp. 537–544.

[75] P. K. JHA AND G. SLUTZKI, *Independence numbers of product graphs*, Appl. Math. Lett., 7 (1994), pp. 91–94.

[76] S. KLAVŽAR, *Some new bounds and exact results on the independence number of Cartesian product graphs*, Ars Combin., 74 (2005), pp. 173–186.

[77] S. KLAVŽAR, *On the canonical metric representation, average distance, and partial Hamming graphs*, European J. Combin., 27 (2006), pp. 68–73.

[78] S. KLAVŽAR, A. LIPOVEC, AND M. PETKOVŠEK, *On subgraphs of Cartesian product graphs*, Discrete Math., 244 (2002), pp. 223–230.

[79] S. KLAVŽAR, U. MILUTINOVIĆ, AND C. PETR, *On the Frame-Stewart algorithm for the multi-peg Tower of Hanoi problem*, Discrete Appl. Math., 120 (2002), pp. 141–157.

[80] S. KLAVŽAR AND I. PETERIN, *Characterizing subgraphs of Hamming graphs*, J. Graph Theory, 49 (2005), pp. 302–312.

[81] S. KLAVŽAR, T.-L. WONG, AND X. ZHU, *Distinguishing labellings of group action on vector spaces and graphs*, J. Algebra, 303 (2006), pp. 626–641.

[82] M. KLEŠČ, *The crossing number of $K_{2,3} \times C_3$*, Discrete Math., 251 (2002), pp. 109–117. Cycles and colourings (Stará Lesná, 1999).

[83] M. KLEŠČ, *Some crossing numbers of products of cycles*, Discuss. Math. Graph Theory, 25 (2005), pp. 197–210.

[84] M. KLEŠČ, R. B. RICHTER, AND I. STOBERT, *The crossing number of $C_5 \times C_n$*, J. Graph Theory, 22 (1996), pp. 239–243.

[85] R. H. LAMPREY AND B. H. BARNES, *A new concept of primeness in graphs*, Networks, 11 (1981), pp. 279–284.

[86] R. H. LAMPREY AND B. H. BARNES, *A characterization of Cartesian-quasiprime graphs*, Congr. Numer., 109 (1995), pp. 117–121.

[87] S. LANG, *Algebra*, vol. 211 of Graduate Texts in Mathematics, Springer-Verlag, New York, third ed., 2002.

[88] B. LIOUVILLE, *Sur la connectivité des produits de graphes*, C. R. Acad. Sci. Paris Sér. A-B, 286 (1978), pp. A363–A365.

[89] E. S. MAHMOODIAN, *On edge-colorability of Cartesian products of graphs*, Canad. Math. Bull., 24 (1981), pp. 107–108.

[90] W. McCuaig and B. Shepherd, *Domination in graphs with minimum degree two*, J. Graph Theory, 13 (1989), pp. 749-762.

[91] A. Meir and J. W. Moon, *Relations between packing and covering numbers of a tree*, Pacific J. Math., 61 (1975), pp. 225-233.

[92] D. J. Miller, *The automorphism group of a product of graphs*, Proc. Amer. Math. Soc., 25 (1970), pp. 24-28.

[93] B. Mohar and C. Thomassen, *Graphs on Surfaces*, Johns Hopkins Studies in the Mathematical Sciences, Johns Hopkins University Press, Baltimore, MD, 2001.

[94] H. M. Mulder, *The Interval Function of a Graph*, vol. 132 of Mathematical Centre Tracts, Mathematisch Centrum, Amsterdam, 1980.

[95] T. Nakayama and J. Hashimoto, *On a problem of G. Birkhoff*, Proc. Amer. Math. Soc., 1 (1950), pp. 141-142.

[96] O. Ore, *Theory of Graphs*, American Mathematical Society Colloquium Publications, Vol. XXXVIII, American Mathematical Society, Providence, R.I., 1962.

[97] E. M. Palmer, *Prime line-graphs*, Nanta Math., 6 (1973), pp. 75-76.

[98] Y. H. Peng and Y. C. Yiew, *The crossing number of $P(3,1) \times P_n$*, Discrete Math., 306 (2006), pp. 1941-1946.

[99] B. Reed, *Paths, stars and the number three*, Combin. Probab. Comput., 5 (1996), pp. 277-295.

[100] R. B. Richter and C. Thomassen, *Intersections of curve systems and the crossing number of $C_5 \times C_5$*, Discrete Comput. Geom., 13 (1995), pp. 149-159.

[101] R. D. Ringeisen and L. W. Beineke, *The crossing number of $C_3 \times C_n$*, J. Combin. Theory Ser. B, 24 (1978), pp. 134-136.

[102] M. Rosenfeld and D. Barnette, *Hamiltonian circuits in certain prisms*, Discrete Math., 5 (1973), pp. 389-394.

[103] G. Sabidussi, *Graphs with given group and given graph-theoretical properties*, Canad. J. Math., 9 (1957), pp. 515-525.

[104] G. Sabidussi, *Graph multiplication*, Math. Z., 72 (1959/1960), pp. 446-457.

[105] C. Schwarz and D. S. Troxell, *$L(2,1)$-labelings of Cartesian products of two cycles*, Discrete Appl. Math., 154 (2006), pp. 1522-1540.

[106] S. Špacapan, personal communication, (2007).

[107] S. ŠPACAPAN, *Connectivity of Cartesian products of graphs*, Appl. Math. Lett., 21 (2008), pp. 682-685.

[108] C. TARDIF, *A fixed box theorem for the Cartesian product of graphs and metric spaces*, Discrete Math., 171 (1997), pp. 237-248.

[109] W. T. TUTTE, *A theorem on planar graphs*, Trans. Amer. Math. Soc., 82 (1956), pp. 99-116.

[110] A. VINCE, *Star chromatic number*, J. Graph Theory, 12 (1988), pp. 551-559.

[111] V. G. VIZING, *The Cartesian product of graphs*, Vyčisl. Sistemy, 9 (1963), pp. 30-43.

[112] V. G. VIZING, *On an estimate of the chromatic class of a p-graph*, Diskret. Anal., 3 (1964), pp. 25-30.

[113] V. G. VIZING, *Some unsolved problems in graph theory*, Uspehi Mat. Nauk, 23 (1968), pp. 117-134.

[114] V. G. VIZING, *Coloring the vertices of a graph in prescribed colors*, Diskret. Analiz, (1976), pp. 3-10, 101.

[115] D. B. WEST, *Introduction to Graph Theory, Second Edition*, Prentice Hall Inc., Upper Saddle River, NJ, 2001.

[116] H. WHITNEY, *Congruent graphs and the connectivity of graphs*, Amer. J. Math., 54 (1932), pp. 150-168.

[117] H. WIENER, *Structural determination of paraffin boiling points*, J. Amer. Chem. Soc., 69 (1947), pp. 17-20.

[118] P. M. WINKLER, *Isometric embedding in products of complete graphs*, Discrete Appl. Math., 7 (1984), pp. 221-225.

[119] J.-M. XU AND C. YANG, *Connectivity of Cartesian product graphs*, Discrete Math., 306 (2006), pp. 159-165.

[120] Y. N. YEH AND I. GUTMAN, *On the sum of all distances in composite graphs*, Discrete Math., 135 (1994), pp. 359-365.

[121] X. ZHU, *Star chromatic numbers and products of graphs*, J. Graph Theory, 16 (1992), pp. 557-569.

[122] X. ZHU, *Circular chromatic number: a survey*, Discrete Math., 229 (2001), pp. 371-410.

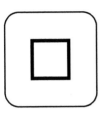

Name Index

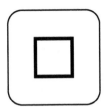

Symbol Index

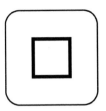

Subject Index